SIMPLIFIED DESIGN OF CONCRETE STRUCTURES

SIMPLIFIED DESIGN OF CONCRETE STRUCTURES

Eighth Edition

JAMES AMBROSE
University of Southern California

and

PATRICK TRIPENY
University of Utah

JOHN WILEY & SONS, INC.

Published by John Wiley & Sons, Inc., Hoboken, New Jersey

Published simultaneously in Canada

For general information about our other products and services, please contact our Customer Care Department within the United States at (800) 762-2974, outside the United States at (317) 572-3993 or fax (317) 572-4002.

Wiley also publishes its books in a variety of electronic formats. Some content that appears in print may not be available in electronic books. For more information about Wiley products, visit our web site at www.wiley.com.

Library of Congress Cataloging-in-Publication Data:

Ambrose, James E.
   Simplified design of concrete structures / James Ambrose with Patrick Tripeny.—8th ed.
     p. cm.
   Includes bibliographical references and index.
   ISBN-13: 978-0-470-04414-8 (cloth)
    1. Reinforced concrete construction.   2. Structural design.   I. Tripeny, Patrick.
II. Title.
   TA683.2.A524   2007
   693'.54—dc22   2006027771

Printed in Mexico

10 9 8 7 6 5 4

CONTENTS

PREFACE TO THE EIGHTH EDITION

Publication of this book presents an opportunity for yet another generation of students of building design to access the subject of concrete structures. The particular focus of this work is a concentration on widely used, simple, and ordinary methods of construction. In addition, the effort has been made to keep mathematical work at a low level, in order to emphasize the accessibility of the work to untrained persons.

The basic purpose of this "simplified" work is well expressed in the preface to the first edition by the originator of the simplified series of books, the late Professor Harry Parker of the University of Pennsylvania. Excerpts from Professor Parker's preface to the first edition follow this preface. To the extent possible, we have adhered to the spirit expressed by Professor Parker.

Of course, structural engineering is no longer really simple. Utilization of all of the available resources for investigation and design of structures is a complex and exhaustive task. Professional engineers must climb this notable learning curve in pursuit of credibility as professionals. We do not mean to belittle the work of serious engineers by our simplified approach. Still, practical means for construction and the use of very ordinary structural products provide the resource for the vast amount of building construction—even in the computer age. It is possible, therefore, to present real structural solutions for ordinary structural tasks with a minimum of complexity. And it is also possible to present the design process for these solutions in relatively elementary form.

Readers of this book should obtain a useful overview of the field of concrete structures and of the means for their design. Those wishing to pursue the study to more advanced levels can find many publications and educational opportunities for their study. For many, seeking mostly only a general view of the topic and an understanding of basic design processes, this book may suffice well.

This topic is supported by an amazing archive of reference materials, including publications and computer programs. We have used a few essential sources for this work, a primary one being the ACI Code, published by the American Concrete Institute. To the extent possible, the work presented here conforms to the specifications of the latest edition of that work. We are grateful to the ACI for permission to use some materials from the code for this work.

Preparation of instructional material needs to include the testing of the materials in classroom situations. The authors of this book have had extensive opportunities to utilize classroom experience for development of the materials in this book. We are considerably in debt to the many students who have sat in our classes and from whom we have undoubtedly learned more than they have from our efforts. We are also indebted to the schools that have provided our teaching opportunities, especially the University of Utah and the University of Southern California. We are very grateful for the support and encouragement provided by these schools to classroom teachers.

We are very appreciative for the support of our publishers, John Wiley and Sons. This publisher has a long history of maintaining a strong catalog of publications in the fields of architecture and construction, and we thank them for that continuing effort. We are especially grateful to our publisher, Amanda Miller, our editor, Paul Drougas, and our production liaison, Nancy Cintron. This book would indeed not have been produced without their significant contributions and support.

Finally, we need to express the gratitude we have to our families. Writing work, especially when added to an already full time occupation, is very time consuming. We thank our spouses and children for their patience, endurance, support, and encouragement in permitting us to achieve this work.

JAMES AMBROSE

PATRICK TRIPENY

PREFACE TO THE
FIRST EDITION

(The following is an excerpt from Professor Parker's preface to the first edition.)

The preparation of this book has been prompted by the fact that many young men desirous of the ability to design elementary reinforced concrete structural members have been deprived of the usual preliminary training. The author has endeavored to simplify the subject matter for those having a minimum of preparation. Throughout the text will be found references to Section 1 of *Simplified Engineering for Architects and Builders*. Familiarity with this brief treatment of the principles of mechanics is sufficient. Any textbook on mechanics will give the desired information. This particular book has been referred to as a convenience in having a direct reference. With these basic principles, and a high school knowledge of algebra, no other preparation is needed.

In preparing material for this book, the author has had in mind its use as a textbook as well as a book to be used for home study. Simple, concise explanations of the design of the most common structural elements have been given rather than discussions of the more involved problems. In addition to the usual design formulas sufficient theory underlying the principles of design is presented, in developing basic formulas, to ensure the student a thorough knowledge of the fundamentals involved.

A major portion of the book contains illustrative examples giving the solution of the design of structural members. Accompanying the examples are problems to be solved by the student.

The author has made no attempt to offer short-cuts or originality in design. Instead, he has endeavored to present clearly and concisely the present day methods commonly used in the design of reinforced concrete members. A thorough knowledge of the principles herein set forth should encourage the student and serve as adequate preparation for advanced study.

HARRY PARKER
High Hollow, Southampton, PA
March 1943

INTRODUCTION

This book deals with the design of concrete structures for buildings. The term *concrete* refers to various forms of materials composed of inert, loose objects held together by a binding matrix. Examples are bituminous concrete (asphalt paving) and lightweight fiber concretes used for roof tiles. In this book, the term refers to the material commonly used for sidewalks and foundations that consists of portland cement, sand, and stone, mixed with water and allowed to harden.

The common forms of concrete, while strong in compression resistance, have a structural flaw; they have low resistance to tension—usually 20% or less of their resistance to compression. For structural applications, common methods for overcoming this limitation are:

- Limiting of concrete's use to applications primarily involving compression, such as columns, piers, and well-supported pavements
- Adding fiber materials to the mix to enhance tension resistance of the basic material

1

- Placing steel bars within the cast concrete at known points of required tension resistance (called *reinforced concrete*)
- Inducing compensating compression forces before structural loads are applied so that the net effect is a significant reduction of internal tension in the concrete (called *prestressing*)

Although all of these methods are described in this book, the design work presented deals primarily with construction of ordinary reinforced concrete.

Structural design is the set of decisions, inventions, and plans that results in a fully described structure, ready for construction. To get to that fully described structure in a practical manner is the designer's pragmatic task. To achieve that task, designers typically use information from previous designs, from observations of previously built structures, from results of research, and from the general body of publications that record the collective experience of concrete construction. Invention, innovation, and experimentation help to advance knowledge, but experience provides the confidence to trust our design practices. This book deals with common usages, reasonably protected by the experiences of many designers and builders.

The materials in this book are not intended for well-trained, experienced structural engineers but rather for people who are interested in the topic but lack both training and experience in structural design. With this readership in mind, the computational work here is reduced to a minimum, using mostly simple mathematical procedures. A minimum background for the reader is assumed in fundamentals of structural mechanics, such as that provided in *Simplified Mechanics and Strength of Materials* (Reference 7) or any other text in basic principles of structural behaviors.

The design work presented here conforms in general with the 2004 edition of *Building Code Requirements for Structural Concrete*, ACI 318–05 (Reference 1), which is published by the American Concrete Institute and is commonly referred to as the ACI Code. For general reference, information is used from *Minimum Design Loads for Buildings and Other Structures*, SEI/ASCE 7–02 (Reference 2), published by the American Society of Civil Engineers, and from the 1997 edition of the *Uniform Building Code, Volume 2: Structural Engineering Provisions* (Reference 3), published by the International Conference of Building Officials. For any actual design work, however, the reader is cautioned to use the codes currently in force in the location of the construction.

TABLE 1   Units of Measurement: U.S. System

Name of Unit	Abbreviation	Use in Building Design
Length		
Foot	ft	Large dimensions, building plans, beam spans
Inch	in.	Small dimensions, size of member cross sections
Area		
Square feet	ft^2	Large areas
Square inches	in.^{2}	Small areas, properties of cross sections
Volume		
Cubic yards	yd^3	Large volumes of soil or concrete (commonly called simply *yards*)
Cubic feet	ft^3	Quantities of materials
Cubic inches	in.^{3}	Small volumes
Force, Mass		
Pound	lb	Specific weight, force, load
Kip	kip, k	1000 pounds
Ton	ton	2000 pounds
Pounds per foot	lb/ft, plf	Linear load (as on a beam)
Kips per foot	kips/ft, klf	Linear load (as on a beam)
Pounds per square foot	lb/ft^2, psf	Distributed load on a surface, pressure
Kips per square foot	k/ft^2, ksf	Distributed load on a surface, pressure
Pounds per cubic foot	lb/ft^3	Relative density, unit weight
Moment		
Foot-pounds	ft-lb	Rotational or bending moment
Inch-pounds	in.-lb	Rotational or bending moment
Kip-feet	kip-ft	Rotational or bending moment
Kip-inches	kip-in.	Rotational or bending moment
Stress		
Pounds per square foot	lb/ft^2, psf	Soil pressure
Pounds per square inch	lb/in.^{2}, psi	Stresses in structures
Kips per square foot	kips/ft^2, ksf	Soil pressure
Kips per square inch	kips/in.^{2}, ksi	Stresses in structures
Temperature		
Degree Fahrenheit	°F	Temperature

Units of Measurement

In this book work is presented basically using U.S. units (feet, inches, etc.). However, for some work equivalent values are indicated for metric units. The form for this double unit presentation is as follows: 246 kip-ft [334 kN-m].

Table 1 (see page 3) lists standard units of measurement in the U.S. system with the abbreviations used in this book and a description of the

TABLE 2   Units of Measurement: SI System

Name of Unit	Abbreviation	Use in Building Design
Length		
Meter	m	Large dimensions, building plans, beam spans
Millimeter	mm	Small dimensions, size of member cross sections
Area		
Square meters	m^2	Large areas
Square millimeters	mm^2	Small areas, properties of member cross sections
Volume		
Cubic meters	m^3	Large volumes
Cubic millimeters	mm^3	Small volumes
Mass		
Kilogram	kg	Mass of material (equivalent to weight in U.S. units)
Kilograms per cubic meter	kg/m^3	Density (unit weight)
Force, Load		
Newton	N	Force or load on structure
Kilonewton	kN	1000 Newtons
Stress		
Pascal	Pa	Stress or pressure (1 pascal = $1 N/m^2$)
Kilopascal	kPa	1000 pascals
Megapascal	MPa	1,000,000 pascals
Gigapascal	GPa	1,000,000,000 pascals
Temperature		
Degree Celsius	°C	Temperature

TABLE 3   Factors for Conversion of Units

To Convert from U.S. Units to SI Units, Multiply by:	U.S. Unit	SI Unit	To Convert from SI Units to U.S. Units, Multiply by:
25.4	in.	mm	0.03937
0.3048	ft	m	3.281
645.2	in.^{2}	mm^2	1.550×10^{-3}
16.39×10^3	in.^{3}	mm^3	61.02×10^{-6}
416.2×10^3	in.^{4}	mm^4	2.403×10^{-6}
0.09290	ft^2	m^2	10.76
0.02832	ft^3	m^3	35.31
0.4536	lb (mass)	kg	2.205
4.448	lb (force)	N	0.2248
4.448	kip (force)	kN	0.2248
1.356	ft-lb (moment)	N-m	0.7376
1.356	kip-ft (moment)	kN-m	0.7376
16.0185	lb/ft^3 (density)	kg/m^3	0.06243
14.59	lb/ft (load)	N/m	0.06853
14.59	kip/ft (load)	kN/m	0.06853
6.895	psi (stress)	kPa	0.1450
6.895	ksi (stress)	MPa	0.1450
0.04788	psf (load or pressure)	kPa	20.93
47.88	ksf (load or pressure)	kPa	0.02093
$0.566 \times (°F - 32)$	°F	°C	$(1.8 \times °C) + 32$

applications for the units. In similar form, Table 2 gives the corresponding units in the metric system. Table 3 lists conversion factors for shifting from one system to the other.

COMPUTATIONAL ACCURACY

Structures for buildings—especially those of sitecast concrete—are seldom produced with a high degree of dimensional precision. Exact dimensions of concrete elements and exact locations of steel reinforcing bars are difficult to achieve, even for the most diligent workers and builders. In addition, the lack of precision in predicting loads for any structure makes highly precise structural computations moot. As a result, numbers are routinely rounded off to the first three digits. This is not a free pass for sloppy computations, but merely a reality for significance in precision of design work.

SYMBOLS

The following symbols are frequently used:

Symbol	Reading
>	is greater than
<	is less than
≥	equal to or greater than
≤	equal to or less than
6′	6 feet
6″	6 inches
Σ	sum of
ΔL	change in L

NOMENCLATURE

Notation used in this book complies generally with that used in the 2005 edition of the ACI Code (Reference 1) and other widely used industry references. The following list contains notation used frequently in this book:

A_c = area of concrete

A_g = gross area of a section, defined by its outer dimensions

A_n = net area

A_s = area of reinforcement (steel bars)

A'_s = area of compressive reinforcement in a doubly reinforced beam

A_v = area of shear reinforcement

C = compressive force

E_c = modulus of elasticity of concrete

E_s = modulus of elasticity of steel

I = moment of inertia

K = slenderness modification factor for a column

L = length (usually of span)

L_d = length of embedment in concrete required for development of stress in a steel bar

L_{dh} = length of embedment in concrete required for development of stress in a steel bar with a hooked end

M = bending moment

M_r = ultimate resisting moment capacity of a section

P = concentrated load or axial column load

S = section modulus

T = tension force

W = 1. total gravity load; 2. weight or dead load of an object; 3. total wind load; 4. total uniformly distributed load or pressure

a = depth of equivalent rectangular stress block in a beam

b = width

b_f = width of flange for a T-beam

b_w = width of stem for a T-beam

c = in bending, distance from extreme fiber stress to neutral axis

d = depth; effective depth for a concrete beam

e = eccentricity

f'_c = specified ultimate compressive strength of concrete

f_y = yield strength of steel

h = height

jd = length of internal moment arm in a beam

l = length

n = modular ratio of moments of inertia: E_s/E_c

p = percent of reinforcement, expressed as A_s/A_g

s = spacing, center-to-center, of reinforcing bars

t = thickness

v = unit shear stress

w = unit of uniformly distributed load per foot of span

ϕ = resistance or strength reduction factor in strength design (the RF in LRFD)

WORK IN THIS BOOK

The work in this book is of limited complexity. This is achieved by presenting the simplest and most common problems and by use of simplified mathematical computations. Although structural computations are critical for responsible structural design, they in fact constitute a small part of the total effort in structural design work. Planning the layout of the structure, integrating the structure with other components of the building construction, developing of construction details, and writing structural specifications are processes that draw on knowledge and skills beyond mathematics and applied physics.

Structural designers employ several forms of computations throughout the design process, including:

- Rough approximations for very preliminary design
- Quick computations, hand done, for more detailed preliminary design
- Round number sizes and product choices from tabulations
- Serious computations with well-established data for detailed preliminary design and cost estimates
- Serious final computations, usually using computer methods
- And possibly—final computations performed by hired consultants with more experience, better computer programs, or highly developed specialized skills

Which methods are used depends on the designers' experience, confidence, resources, and the time allowed for the computational work. On large design projects, all of these methods are commonly used for specific needs at various stages of design.

Computations in this book are done mostly in the solutions of example problems. Study opportunities for the reader are presented in exercise problems similar to the demonstrated problems in the examples. Answers for exercise problems are provided at the back of the book.

REFERENCE SOURCES

Most of the data and reference materials for the work in this book are provided here. However, the materials presented here are typically a small fraction of available source materials. For expanded study opportunities, readers will benefit from gaining some familiarity with the general references for this topic—most notably with the ACI Code (Reference 1).

Also useful for expanded study is a current text for the subject intended for students in civil engineering courses. Such a text is that listed as Reference 4 for this book.

A major type of source for design work is that obtainable from various industry and professional organizations. Some of these organizations related to work in design of concrete structures are the following:

American Concrete Institute. This organization has a wide-ranging membership that includes major companies, as well as many outstanding engineers, teachers, and researchers. Its major publication is the ACI Code.

Portland Cement Association. This organization is primarily sponsored by producers of cement. It conducts important research and is a major producer of publications.

Concrete Reinforcing Steel Institute. Although steel reinforcement constitutes a small volume in reinforced concrete, its cost is a major consideration for construction. The CRSI sponsors research and produces many publications.

Prestressed Concrete Institute. This organization publishes materials relating to industrially produced concrete structural elements, as well as general uses of prestressing and precasting.

National Concrete Masonry Association. Because masonry construction and design tend to reflect regional concerns, many masonry industry organizations exist. The NCMA is a national organization that seeks to coordinate standards and practices on a national basis.

Masonry Institute of America. This organization has considerable influences on design and construction of reinforced masonry—a system widely used in earthquake-prone and windstorm-prone regions (principally southern and western parts of the United States).

1

STRUCTURAL USE
OF CONCRETE

This chapter presents some of the considerations for the use of concrete for structural purposes in building construction.

1.1 CONCRETE AS A STRUCTURAL MATERIAL

Concrete consists of a mixture that contains a mass of loose, inert particles of graded size (commonly sand and gravel) held together in solid form by a binding agent. That general description covers a wide range of end products. The loose particles may consist of wood chips, industrial wastes, mineral fibers, and various synthetic materials. The binding agent may be coal tar, gypsum, portland cement, or various synthetic compounds. The end products range from asphalt pavement, insulating fill, shingles, wall panels, and masonry units to the familiar sidewalks, roadways, foundations, and building frameworks.

This book deals primarily with concrete formed with the common binding agent of *portland cement,* and a loose mass consisting of sand

and gravel. This is what most of us mean when we use the term *concrete*. With minor variations, this is the material used mostly for structural concrete—to produce building structures, pavements, and foundations.

Concrete made from natural materials was used by ancient builders thousands of years ago. Modern concrete, made with industrially produced cement, was first produced in the early part of the nineteenth century when the process for producing portland cement was developed. Because of its lack of tensile strength, however, concrete was used principally for crude, massive structures—foundations, bridge piers, and heavy walls.

In the late nineteenth century, several builders experimented with the technique of inserting iron or steel rods into relatively thin structures of concrete to enhance their ability to resist tensile forces. This was the beginning of what we now know as *reinforced concrete*. Many of the basic forms of construction developed by these early experimenters have endured to become part of our common technical inventory for building structures.

Over the years, from ancient times until now, there has been a steady accumulation of experience derived from experiments, research, and, most recently, intense development of commercial products. As a result, there is currently available to the building designer an immense variety of products under the general classification of concrete. This range is somewhat smaller if major structural usage is required, but the potential variety is still significant.

1.2   COMMON FORMS OF CONCRETE STRUCTURES

For building structures, concrete is mostly used with one of three basic construction methods. The first is called *sitecast concrete,* in which the wet concrete mix is deposited in forms at the location where it is to be used. This method is also described as *cast-in-place* or *in situ* construction.

A second method consists of casting portions of the structure at a location away from the desired location of the construction. These elements—described as *precast concrete*—are then moved into position, much as blocks of stone or parts of steel frames are.

Finally, concrete may be used for masonry construction—in one of two ways. Precast units of concrete called concrete masonry units (CMUs), may be used in a manner similar to bricks or stones. Alternately, concrete fill may be used to produce solid masonry by being poured into cavities in masonry produced with bricks, stone, or CMUs. The latter technique, combined with the insertion of steel reinforcement into the cavities, is widely used for masonry structures today. The use of concrete-filled masonry, however, is one of the oldest forms of concrete construction—used extensively by the Romans and the builders of early Christian churches.

Concrete is produced in great volume for various forms of construction. Building frames, walls, and other structural systems represent a minor usage of the total concrete produced. Pavements for sidewalks, parking lots, streets, and ground-level floor slabs in buildings use more concrete than all the building frameworks. Add the usage for the interstate highway system, water control, marine structures, and large bridges and tunnels, and building structural usage shrinks considerably in significance. One needs to understand this when considering the economics and operations of the concrete industry.

Other than pavements, the widest general use of concrete for building construction is foundations. Almost every building has a concrete foundation, whether the major above ground construction is concrete, masonry, wood, steel, aluminum, or fabric. For small buildings with shallow footings and no basement, the total foundation system may be modest, but for large buildings and those with many belowground levels, there may well be a gigantic underground concrete structure.

For above ground building construction, concrete is generally used in situations that fully realize the various advantages of the basic material and the common systems that derive from it. For structural applications, this means using the major compressive resistance of the material and in some situations its relatively high stiffness and inertial resistance (major dead weight). However, in many applications, the nonrotting, vermin- and insect-resistive, and fire-resistive properties may be of major significance. And for many uses, its relatively low bulk-volume cost is important.

Elements of Concrete Structures

Formation of a concrete structural system for a building usually consists of the assemblage of individual structural elements. Most commonly used structural systems are combinations of a few basic elements; these are:

- Structural walls
- Structural columns, piers, or other single supports
- Horizontal-spanning beams
- Horizontal-spanning decks

The actions of these individual elements and their various interactions for structural functions must be considered when designing building structures. Concrete is also widely used for foundations, and the common elements utilized for this purpose are:

- Foundation walls
- Wall and single-column-bearing footings
- Pile caps for clusters of piles
- Piers, cast as columns in excavated holes

Consideration is given to each of these individual elements in this book. Some of the possibilities for their use in whole, assembled structures are illustrated in the building case study examples in Chapter 16.

Many special elements are also typically required for the completion of any building structure, such as pilasters, brackets, keys, pedestals, column caps, and so on. These are necessary, but essentially secondary, elements of the basic systems. Various situations for their use are illustrated in this book.

Many structures of more exotic forms can be realized with concrete beyond the simple systems treated in this book. Arches, domes, thin shells, folded plates, and other imaginative systems have been developed by designers who push the limits of the material's potentialities. We hope that readers may have the opportunity to work on such exciting structures at some time. Here, we start with the simplest, and most commonly used, structures.

1.3 PRIMARY SITUATIONS FOR INVESTIGATION AND DESIGN

A critical step in the visualization of structural behaviors is the consideration of the basic internal structural actions that occur in structural members. The five primary actions of internal structural resistance are tension, compression, shear, bending, and torsion. The structural functions of all the basic elements described previously can be developed with combinations of these basic internal actions.

There is another level down, of course, consisting of the basic stress actions that are a material's direct response to structural forces. Thus all the internal force actions can be produced from the basic stresses of tension, compression, and shear. For some materials the character of the stress is a critical concern since the material responds differently to the different stresses. Such is indeed the case with concrete, for which development of tension stress is a problem; this is the starting point for the design of reinforcement.

For our purposes here, it is useful to start with a basic element: the beam. This immediately presents all three basic stresses in the development of bending and shear for the basic beam action. And it makes a case for reinforcement, to develop significant internal tension for bending resistance, as well as an enhanced resistance to shear. The spanning slab represents essentially a variation on the basic beam function.

The second basic element to be considered is the column; that is the element whose basic task is resistance to compression. Variations here consist of the pier or pedestal (a very short column) and the bearing wall.

Finally, for the assembled system, a significant consideration is the interaction of elements in various framed configurations. This introduces the problem of joints or connections between elements, with the various force transfers necessary through the joints. It also involves consideration of the effects of one element on others to which it is connected; for example, the actions of adjacent beams on each other when continuity between spans occurs, and the interactions of beams and columns in a planar frame with continuous elements.

1.4 MATERIALS AND NATURE OF STRUCTURAL CONCRETE

This section presents discussions of the various ingredients of structural concrete and factors that influence the physical properties of the finished concrete. Other elements used to produce concrete structures are also discussed.

Common Forms of Structural Concrete

For serious structural usage, concrete must attain significant strength and stiffness, reasonable surface hardness, and other desired properties. While the mixture used to obtain concrete can be almost endlessly varied, the controlled mixes used for structural applications are developed within a quite limited set of variables. The most commonly used mix contains ordinary portland cement, clean water, medium-to-coarse sand, and a considerable volume of some fairly large pieces of rock. This common form of concrete will be used as a basis for comparison of mixes for special purposes.

Figure 1.1 shows the composition of ordinary concrete. The binder consists of the water and cement, whose chemical reaction results in the

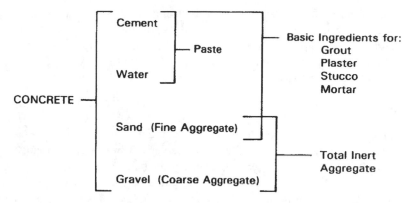

Figure 1.1 Composition of ordinary concrete.

hardening of the mass. The binder is mixed with some aggregate (loose, inert particles) so that the binder coats the surfaces and fills the voids between the particles of the aggregate. For materials such as grout, plaster, and stucco, the aggregate consists of sand of reasonably fine grain size. For concrete the grain size is extended into the category of gravel, with the maximum particle size limited only by the size of the structure. The end product—the hardened concrete—is highly variable, due to the choices for the individual basic ingredients; modifications in the mixing, handling, and curing processes; and possible addition of special ingredients.

Cement

The cement used most extensively in building construction is portland cement. Of the five standard types of portland cement generally available in the United States and for which the American Society for Testing and Materials has established specifications, two types account for most of the cement used in buildings. These are a general-purpose cement for use in concrete designed to reach its required strength in about 28 days, and a high-early-strength cement for use in concrete that attains its design strength in a period of a week or less.

All portland cements set and harden by reacting with water, and this hydration process is accompanied by generation of heat. In massive concrete structures such as dams, the resulting temperature rise of the materials becomes a critical factor in both design and construction, but the problem is usually not significant in building construction. A low-heat cement is designed for use where the heat rise during hydration is a critical factor. It is, of course, essential that the cement actually used in construction correspond to that employed in designing the mix, to produce the specified compressive strength of the concrete.

Mixing Water

Water must be reasonably clean, free of oil, organic matter, and any substances that may affect the actions of hardening, curing, or general finish quality of the concrete. In general, drinking-quality (potable) water is usually adequate. Salt-bearing seawater may be used for plain concrete (without reinforcing) but may cause the corrosion of steel bars in reinforced concrete.

A critical concern for the production of good concrete is the *amount* of water used. In this regard there are three principal concerns, as follows:

1. Having enough water to react chemically with the cement so that the hardening and strength gain of the concrete proceeds over time until the desired quality of material is attained.
2. Having enough water to facilitate good mixing of the ingredients and allow for handling in casting and finishing of the concrete.
3. Having the amount of water low enough so that the combination of water and cement (the paste) is not too low in cement to perform its bonding action. This is a major factor in producing high-grade concrete for structural applications.

Stone Aggregate

The most common aggregates are sand, crushed stone, and pebbles. Particles smaller than 3/16 in. in diameter constitute the *fine aggregate*. There should be only a very small amount of very fine materials, to allow for the free flow of the water-cement mixture between the aggregate particles. Material larger than 3/16 in. is called the *coarse aggregate*. The maximum size of the aggregate particle is limited by specification, based on the thickness of the cast elements, spacing and cover of the reinforcing, and some consideration of finishing methods.

In general, the aggregate should be well graded, with some portion of large to small particles over a range to permit the smaller particles to fill the spaces between the larger ones. The volume of the concrete is, thus, mostly composed of the total aggregate, the water and cement going into the spaces remaining between the smallest aggregate particles. The weight of the concrete is determined largely by the weight of the coarse aggregate. Strength is also dependent, to some degree, on the structural integrity of the large aggregate particles.

Special Aggregates

While stone is the most common coarse aggregate, for various reasons other materials may be used. One reason for this may be the absence of available stone of adequate quality, but more often there is some desire to impart particular modified properties to the concrete. Some of these desired properties and the types of aggregates used to achieve them are discussed in this section.

Weight Reduction.   For structural concrete, a common desire is for some reduction of the dead load of the structure. This is most often desired for concrete elements of spanning structures. Since the coarse aggregate typically constitutes at least two-thirds of the total mass of the concrete, any significant reduction in unit density of the coarse aggregate will result in a significant weight reduction of the finished concrete. If a relatively high strength is also desired, there is a limit to how much reduction can be achieved. Various natural and synthetic materials may be used as substitutes for the ordinary stone, but if reasonable strength and stiffness is critical, the maximum reduction is usually around 25 to 30%; that is, a reduction from a typical density of 145 to 150 pounds per cubic foot (pcf) to something just over 100 pcf. Lower finished densities may be achieved, but usually with significant loss of both strength and stiffness.

Weight Increase.   In some circumstances an increase of weight may be desired. This is usually achieved with selected stone of high density, typically one containing metal ores. Alternately, in some cases, it can be accomplished by using scrap iron as part of the aggregate. This seldom involves structures that are exposed to air or to view.

Better Resistance to Fire.   Individual types of stone have different actions when exposed to the extreme heat of fires. This action may be critical for the concrete structure itself or for its utilization in providing fireproofing for a steel structure. A specific material may be selected for this property and may be either a natural stone or a synthetic product.

Fiber Aggregate. Fibrous materials may be added to the concrete in significant amounts, usually for the increased tension resistance they provide for the concrete. However, these are not used in amounts that significantly reduce the total mass of coarse aggregate. Thus, the development of a fibrous concrete still involves the selection of some material for the coarse aggregate.

For some uses of concrete, it may be possible to utilize some available material for part of the coarse aggregate to reduce cost or to achieve some goal for utilization of the material. In some coastal areas clam shells have been used as part of the coarse aggregate, usually because of the limited availability of good stone. Crushed, recycled glass has been used in limited amounts for some pavements and foundations. However, when the best structural concrete possible is desired, the choice is still most often for some good type of stone that is locally available in sufficient quantity.

Additions to the Basic Concrete Mix

Substances added to concrete to improve its workability, accelerate its set, harden its surface, and increase its waterproof qualities are known as *admixtures.* The term embraces all materials other than the cement, water, and aggregates that are added just before or during mixing. Many proprietary compounds contain hydrated lime, calcium chloride, and kaolin. Calcium chloride is the most commonly used admixture for accelerating the set of concrete, but corrosion of steel reinforcement may be the consequence of its excessive use.

Air-entrained concrete is produced by using an air-entraining portland cement or by introducing an air-entraining admixture as the concrete is mixed. Air-entraining agents produce billions of microscopic air cells per cubic foot; they are distributed uniformly throughout the mass. These minute voids prevent the accumulation of water, which, on freezing, would expand and result in spalling of the exposed surface under frost action. In addition to improving workability, entrained air permits lower water-cement ratios and significantly improves the durability of hardened concrete.

1.5 SIGNIFICANT PROPERTIES OF CONCRETE

In the production of elements of concrete structures, some particular properties of concrete emerge as most significant. This section discusses these major properties.

Strength

The primary index of strength of concrete is the *specified compressive strength,* designated f'_c. This is the unit compressive stress used for structural design and for a target for the mix design. It is usually given in units of psi, and it is a common practice to refer to the structural quality of the concrete simply by calling it by this number: 3000-lb concrete, for example. For strength design, this value is used to represent the ultimate compressive strength of the concrete. For working stress design, allowable maximum stresses are based on this limit, specified as some fraction of f'_c.

Hardness

The hardness of concrete refers essentially to its surface density. This is dependent primarily on the basic strength, as indicated by the value for compressive stress. However, surfaces may be somewhat softer than the central mass of concrete, owing to early drying at the surface. Some techniques are used to deliberately harden surfaces, especially those of the tops of slabs. Fine troweling will tend to draw a very cement-rich material to the surface, resulting in an enhanced density. Chemical hardeners can also be used, as well as sealing compounds that trap surface water and enhance the natural hardening process of the surface.

Stiffness

Stiffness of structural materials is a measure of resistance to deformation under stress. For compression and tension stress resistance, stiffness is measured by the *modulus of elasticity,* designated E. This modulus is established by tests and is the ratio of unit stress to unit strain. Since unit strain has no unit designation (measured as inch/inch, etc.), the unit for E, thus, becomes the unit for stress, usually lb/in.^{2} [MPa].

The magnitude of elasticity for concrete, E_c, depends on the weight of the concrete and its strength. For values of unit (w) weight between 90 and 155 lb/ft^3 or pcf, the value of E_c is

$$E_c = w^{1.5}33\sqrt{f'_c}$$

The unit weight for ordinary stone-aggregate concrete is usually assumed to be an average of 145 pcf. Substituting this value for w in the equation, we obtain a simpler form for the concrete modulus of

$$E_c = 57,000\sqrt{f'_c}$$

For metric units, with stress measured in MPa, the expression becomes

$$E_c = 4730\sqrt{f'_c}$$

Distribution of stresses and strains in reinforced concrete is dependent on the concrete modulus, the steel modulus being a constant. This is discussed in Chapter 7. In the design of reinforced concrete members we employ the term n. This is the ratio of the modulus of elasticity of steel to that of concrete, or $n = E_s/E_c$. E_s is taken as 29,000 ksi [200,000 MPa], a constant. The value for concrete, however, is variable, as we have seen. Values for n are usually given in tables of properties, although they are typically rounded off.

When subjected to long-duration stress at a high level, concrete has a tendency to *creep*, a phenomenon in which strain increases over time under constant stress. This has effects on deflections and on the distributions of stresses between the concrete and reinforcing. Some of the implications of this for design are discussed in the chapters that deal with beams and columns.

As discussed in other sections, there are various controls that can be exercised to ensure a desired type of material in the form of the hardened concrete. The three properties of greatest concern are the water content of the wet mix and the density and compressive strength of the hardened concrete. Design of the mix, handling of the wet mix, and curing of the concrete after casting are the general means of controlling the end product.

In addition to the basic structural properties, there are various properties of concrete that bear on its use as a construction material and in some cases on its structural integrity.

Workability

This term generally refers to the ability of the wet mixed concrete to be handled, placed in the forms, and finished while still fluid. A certain degree of workability is essential to the proper forming and finishing of the material. However, the fluid nature of the mix is largely determined by the amount of water present, and the easiest way to make it more workable is to add water. Up to a point, this may be acceptable, but the extra water usually means less strength, greater porosity, and more shrinkage— all generally undesirable properties. Use is made of vibration, admixtures, and other techniques to facilitate handling without increasing the water content.

Watertightness

It is usually desirable to have a generally nonporous concrete. This may be quite essential for walls or for floors consisting of paving slabs, but is good in general for protection of reinforcing from corrosion. Watertightness is obtained by having a well-mixed, high-quality concrete (low water content, etc.) that is worked well into the forms and has dense surfaces with little cracking or voids. Concrete is absorptive, however, and when subjected to the continuous presence of water will become saturated. Moisture or waterproof barriers must be used where water penetration must be prevented.

Density

Concrete unit weight is essentially determined by the density of the coarse aggregate (ordinarily two-thirds or more of the total volume) and the amount of air in the mass of the finished concrete. With ordinary gravel aggregate and air limited to not more than 4% of the total volume, air dry concrete weighs around 145 lb/ft^3. Use of strong but lightweight aggregates can result in weight reduction to close to 100 lb/ft^3 with strengths generally competitive with that obtained with gravel. Lower densities are achieved by entraining air up to 20% of the volume and using very light aggregates, but strength and other properties are quickly reduced.

Fire Resistance

Concrete is noncombustible and its insulative, fire protection character is used to protect the steel reinforcing. However, under long exposure to fire, popping and cracking of the material will occur, resulting in actual structural collapse or a diminished capacity that requires replacement or repair after a fire. Design for fire resistance involves the following basic concerns:

- *Thickness of Parts.* Thin walls or slabs may crack quickly, permitting penetration of fire and gases.
- *Cover of Reinforcement.* More insulating protection is required for higher fire rating of the construction.
- *Character of the Aggregate.* Some are more vulnerable to high temperatures.

Design specifications and building code regulations deal with these issues, some of which are discussed in the development of the building design illustrations in Chapter 16.

Shrinkage

Water-mixed materials, such as plaster, mortar, and concrete, tend to shrink during the hardening process. For ordinary concrete, the shrinkage averages about 2% of the volume. Dimensional change of structural members is usually less, due to the presence of the steel bars; however, some consideration must be given to the shrinkage effects. Stresses caused by shrinkage are in some ways similar to those caused by thermal change, the combination resulting in specifications for minimum two-way reinforcing in walls and slabs. For the structure in general, shrinkage is usually dealt with by limiting the size of individual pours of concrete because the major shrinkage ordinarily occurs quite rapidly in the fresh concrete. For special situations, it is possible to modify the concrete with admixtures or special cements that cause a slight expansion to compensate for the normal shrinkage.

1.6 REINFORCEMENT

For most structural applications of concrete, it is necessary to compensate for the weakness of the material in resisting tension. The primary means of accomplishing this is to use steel reinforcing bars. A more recent development is to add fibrous materials to the concrete mix to alter the properties of the basic material.

Steel Reinforcement

The steel used in reinforced concrete consists of round bars, mostly of the deformed type, with lugs or projections on their surfaces. The surface deformations help to develop a greater bond between the steel rods and the enclosing concrete mass. The essential purpose of steel reinforcement is to reduce the cracking of the concrete due to tensile stresses. Structural actions are investigated for the development of tension in the structural members, and steel reinforcement in the proper amount is placed within the concrete mass to resist the tension. In some situations steel reinforcement may also be used to increase compressive resistance since the ratio of magnitudes of strength of the two materials is quite high; thus, the steel displaces a much weaker material and the member gains significant strength.

Tension can also be induced by shrinkage of the concrete during its drying out from the initial wet mix. Temperature variations may also induce tension in many situations. To address these latter occurrences, a minimum amount of reinforcing is used in surface-type members, such as walls and paving slabs, even when no structural action is anticipated.

The most common grades of steel used for ordinary reinforcing bars are Grade 40 and Grade 60, having yield strengths of 40 ksi [276 MPa] and 60 ksi [414 MPa], respectively. The yield strength of the steel is of primary interest for two reasons. Plastic yielding of the steel generally represents the limit of its practical utilization for reinforcing of the concrete because the extensive deformation of the steel in its plastic range results in major cracking of the concrete. Thus, for service load conditions, it is desirable to keep the stress in the steel within its elastic range of behavior where deformation is minimal. (See Figure 1.2.)

The second reason for the importance of the yield character of the reinforcing is its ability to impart a generally yielding nature (plastic deformation character) to the otherwise typically very brittle concrete structure. This is of particular importance for dynamic loading and is a major consideration in design for earthquake forces. Also of importance is the residual strength of the steel beyond its yield stress limit. As shown in

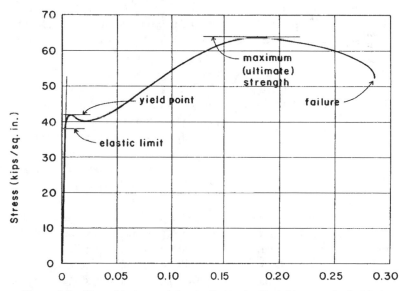

Figure 1.2   Stress/strain graph for ductile steel with yield strength of 40 ksi.

the graph in Figure 1.2, the steel continues to resist stress in its plastic range and then gains a second, higher, strength before failure. Thus, the failure induced by yielding is only a first stage response and a second level of resistance is reserved.

Ample concrete protection, called *cover,* must be provided for the steel reinforcement. This is important to protect the steel from rusting and to be sure that it is well engaged by the mass of concrete. Cover is measured as the distance from the outside face of the concrete to the edge of the reinforcing bar.

Code minimum requirements for cover are ¾ in. for walls and slabs and 1.5 in. for beams and columns. Additional distance of cover is required for extra fire protection or for special conditions where the concrete surface is exposed to weather or is in contact with the ground.

Where multiple bars are used in concrete members (which is the common situation), there are both upper and lower limits for the spacing of the bars. Lower limits are intended to facilitate the flow of wet concrete during casting and to permit adequate development of the concrete-to-steel stress transfers for individual bars. Maximum spacing is generally intended to ensure that there is some steel that relates to a concrete mass of limited size; that is, there is not too extensive a mass of concrete with no reinforcement. For relatively thin walls and slabs, there is also a concern with the scale of spacing related to the thickness of the concrete. Specific code requirements for bar spacing are discussed in Section 2.6.

For structural members, the amount of reinforcement is determined from structural computations as that required for the tension force in the member. This amount (in total cross-sectional area of the steel) is provided by some combination of bars. In various situations, however, there is a minimum amount of reinforcement that is desirable, which may on occasion exceed the amount determined by computation.

Minimum reinforcement may be specified as a minimum number of bars or as a minimum amount, the latter usually based on the cross-sectional area of the concrete member. These requirements are discussed in the sections that deal with the design of the various types of structural members.

In early concrete work, reinforcing bars took various shapes. A problem that emerged was how to properly bond the steel bars within the concrete mass, due to the tendency of the bars to slip or pull out of the concrete. This issue is still a critical one and is discussed in Chapter 8.

In order to anchor the bars in the concrete, various methods were used to produce something other than the usual smooth surfaces on bars. After

TABLE 1.1   Properties of Deformed Reinforcing Bars

Bar Size Designation	Nominal Weight		Nominal Dimensions			
			Diameter		Cross Sectional Area	
	lb/ft	kg/m	in.	mm	in.^{2}	mm^2
No. 3	0.376	0.560	0.375	9.5	0.11	71
No. 4	0.668	0.994	0.500	12.7	0.20	129
No. 5	1.043	1.552	0.625	15.9	0.31	200
No. 6	1.502	2.235	0.750	19.1	0.44	284
No. 7	2.044	3.042	0.875	22.2	0.60	387
No. 8	2.670	3.974	1.000	25.4	0.79	510
No. 9	3.400	5.060	1.128	28.7	1.00	645
No. 10	4.303	6.404	1.270	32.3	1.27	819
No. 11	5.313	7.907	1.410	35.8	1.56	1006
No. 14	7.650	11.390	1.693	43.0	2.25	1452
No. 18	13.600	20.240	2.257	57.3	4.00	2581

much experimentation and testing, a single set of bars was developed with surface deformations consisting of ridges. These deformed bars were produced in graduated sizes with bars identified by a single number (see Table 1.1).

For bars numbered 2 through 8, the cross-sectional area is equivalent to a round bar having a diameter of as many eighths of an inch as the bar number. Thus, a No. 4 bar is equivalent to a round bar of ⁴/₈ or 0.5 in. diameter. Bars numbered from 9 up lose this identity and are essentially identified by the tabulated properties in a reference document.

The bars in Table 1.1 are developed in U.S. units but can, of course, be used with their properties converted to metric units. However, a new set of bars has recently been developed, deriving their properties more logically from metric units. The properties of these bars are given in Table 1.2. The general range of sizes is similar for both sets of bars, and design work can readily be performed with either set. Metric-based bars are obviously more popular outside the United States, but for domestic use (nongovernment) in the United States, the old bars are still in wide use. This is part of a wider conflict over units, which is still going on.

The work in this book uses the old inch-based bars, simply because the computational examples are done in U.S. units. In addition, most of the references still in wide use have data presented basically with U.S. units and the old bar sizes.

TABLE 1.2   Properties of Deformed Reinforcing Bars (Metric Series)

| Bar Size Designation | Nominal Mass (kg/m) | Nominal Dimensions | | Comparison of Area with in. - lb. Bars |
		Diameter (mm)	Cross Sectional Area (mm^2)	
10M	0.785	11.3	100	22% < No. 4
15M	1.570	16.0	200	Same as No. 5
20M	2.355	19.5	300	6% > No. 6
25M	3.925	25.2	500	2% < No. 8
30M	5.495	29.9	700	9% > No. 9
35M	7.850	35.7	1000	1% < No. 11
45M	11.775	43.7	1500	3% > No. 14
55M	19.625	56.4	2500	3% < No. 18

Fiber Reinforcement

Experiments have been conducted over many years on including fibrous elements in the concrete mix with the intention of giving an enhanced tension resistance to the basic hardened concrete. Steel needles, glass, and various mineral fibers have been used. The resulting tensile-enhanced material tends to resist cracking; permit very thin, flexible elements; resist freezing; and permit some applications without steel reinforcing rods. Only minor structural applications have been attempted, but the material is now commonly used for pavements and for thin roof tiles and cladding panels.

1.7   PRESTRESSED CONCRETE

Prestressing consists of the deliberate inducing of some internal stress condition in a structure prior to its sustaining of service loads. The purpose is to compensate in advance for the anticipated service load stress, which for concrete means some high level of tension stress. The "pre-" or "before" stress is, therefore, usually a compressive or reversal bending stress. This section discusses some uses of prestressing and some of the problems encountered in utilizing it for building structures.

Use of Prestressing

Prestressing is principally used for spanning elements, in which the major stress conditions to be counteracted are tension from bending and diagonal tension from shear. A principal advantage of prestressing is that, when properly achieved, it does not result in the natural tension cracking associated with ordinary reinforced concrete. Since flexural cracking is proportionate to the depth of the member, which in turn is proportionate to the span, the use of prestressing frees spanning concrete members from the span limits associated with ordinary reinforcing. Thus, gigantic beam cross sections and phenomenal spans are possible—and indeed have been achieved, although mostly in bridge construction.

The cracking problem also limits the effective use of very high strengths of concrete with ordinary reinforcing. Free of this limit, the prestressed structure can utilize effectively the highest strengths of concrete achievable, and, thus, weight saving is possible, resulting in span-to-weight ratios that partly overcome the usual massiveness of spanning concrete structures.

The advantages just described have their greatest benefit in the development of long, flat-spanning roof structures. Thus, a major use of prestressing has been in the development of precast, prestressed units for roof structures. The hollow-cored slab, single-tee, and double-tee sections shown in Figure 1.3 are the most common forms of such units—now a standard part of our structural inventory. These units can also be used for floor structures, with a major advantage when span requirements are at the upper limits of feasibility for ordinary reinforced construction.

- *For Columns.* Concrete shafts may be prestressed for use as building columns, precast piles, or posts for street lights or signs. In this case the prestressing compensates for bending, shear, and torsion associated with service use and handling during production, transportation, and installation. The ability to use exceptionally high strength concrete is often quite significant in these applications.
- *For Two-Way Spanning Slabs.* Two-way, continuous prestressing can be used to provide for the complex deformations and stress conditions in concrete slabs with two-way spanning actions. A special usage is that for a paving slab designed as a spanning structure where ground settlement is anticipated. Crack reduction may be a significant advantage in these applications.
- *Tiedown Anchors.* When exceptionally high anchorage forces must be developed, and development of ordinary tension reinforcing may be difficult or impossible, it is sometimes possible to use the tension strands employed for prestressing. Large abutments, counterforts for large retaining walls, and other elements requiring considerable tension anchorage are sometimes built as prestressed elements.

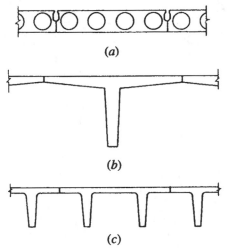

Figure 1.3 Forms of typical products of precast concrete used for horizontal-spanning structures.

- *Horizontal Ties.* Single-span arches and rigid frames that develop outward thrusts on their supports are sometimes tied with prestressing strands.

For any structure it is necessary to consider various loading conditions that occur during construction and over a lifetime of use. For the prestressed structure, this is a quite complex issue, and design must incorporate many different events over the life of the structure. For common usages, experience has produced various empirical adjustments (educated fudge factors) that account for the usual occurrences. For unique applications, there must be some reasonable tolerance for errors in assumptions or some provision for tuning up the finished structure. The prestressed structure is a complex object, and the design of other than very routine elements should be done by persons with considerable training and experience.

Pretensioned Structures

Prestressing is generally achieved by stretching high-strength steel strands (bunched wires) inside the concrete element. The stretching force is eventually transferred to the concrete, producing the desired compression in the concrete. There are two common procedures for achieving the stretching of the strands: pretensioning and post-tensioning.

Pretensioning consists of stretching the strands prior to pouring the concrete. The strands are left exposed in the forms, the concrete is cast around them, and as the concrete hardens, it bonds to the strands. When the concrete is sufficiently hardened, the external stretching force is released, and the strand tension is transferred to the concrete through the bond action on the strand surfaces. This procedure requires some substantial element to develop the necessary resistance to the jacking force used to stretch the strands before the concrete is poured. Pretensioning is used mostly for factory precast units, for which the element resisting the stretching force is the casting form, sturdily built of steel and designed for repeated use.

Pretensioning is done primarily for cost-saving reasons. There is one particular disadvantage to pretensioning: it does not allow for any adjustment, and the precise stress and deformation conditions of the finished product are only approximately predictable. The exact amount of the strand bonding and the exact properties of the finished concrete are somewhat variable. Good quality control in production can keep the range of variability within some bounds, but the lack of precision must be allowed for in the design and construction. A particular problem is the control of deflection of adjacent units in systems consisting of side-by-side units.

Post-Tensioned Structures

In post-tensioned structures, the prestressing strands are installed in a slack condition, typically wrapped with a loose sleeve or conduit. The concrete is poured and allowed to harden around the sleeves and the end anchorage devices for the strand. When the concrete has attained sufficient strength, the strand is anchored at one end and stretched at the other end by jacking against the concrete. When the calibrated jacking force is observed to be sufficient, the jacked end of the strand is locked into the anchorage device, pressurized grout is injected to bond the strand inside the sleeve, and the jack is released.

Post-tensioning is generally used for elements that are cast in place because the forms need not resist the jacking forces. However, it may also be used for precast elements when jacking forces are considerable and/or a higher control of the net existing force is desired.

Until the strands are grouted inside the sleeves, they may be rejacked to a higher stretching force condition repeatedly. In some situations this is done as the construction proceeds, permitting the structure to be adjusted to changing load conditions.

Post-tensioning is usually more difficult and more costly, but there are some situations in which it is the only alternative for achieving the prestressed structure.

1.8 DESIGN OF CONCRETE MIXES

From a structural design point of view, mix design basically means dealing with the considerations involved in achieving a particular design strength, as measured by the value of the fundamental property: f'_c. For this, the principal factors are:

Cement Content

The amount of cement per unit volume is a major factor determining the richness of the cement-water paste and its ability to fully coat all of the aggregate particles and to fill the voids between them. Cement content is normally measured in terms of the number of sacks of cement (1 cubic foot each) per cubic yard of concrete mixed. The average for structural concrete is about 5 sacks per yard. If tests show that the mix exceeds or falls short of the desired results, the cement content is decreased or increased, respectively. The cement is by far the costliest ingredient, so its volume is critical in cost control.

Water-Cement Ratio

This is expressed in terms of gallons of water per sack of cement or gallons of water per yard of mixed concrete. The latter is usually held very close to an average of about 35 gallons per yard; less, and workability is questionable; more, and strength becomes difficult to obtain. Attaining very strong concrete usually means employing various means to reduce water content and to improve the ratio of cement to water without losing workability.

Fineness Modulus of the Sand

If the sand is too coarse, the wet mix will be grainy, and surfaces will be difficult to finish smoothly. If the sand is too fine, an excess of water will be required, resulting in high shrinkage and loss of strength due to thinning out of the water/cement mixture. Grain size and size range are controlled by specification.

Character of the Coarse Aggregate

Shape, size limits, and type of material must be considered. Because this represents the major portion of the concrete volume, its properties are quite important to strength, weight, fire performance, and so on.

Development of the Mix Design

Concrete is obtained primarily from industrial plants that mix the materials and deliver them in mixer trucks. The mix design is developed cooperatively with the structural designer and the management of the mixing plant. Local materials must be used, and experience with them is an important consideration.

1.9   SPECIAL CONCRETES

Within the range of the general material discussed here, there are many special forms of concrete used for special situations and applications. Some of the principal ones are:

Lightweight Structural Concrete

This is concrete that achieves a significant reduction in weight, while retaining sufficient levels of structural properties to remain feasible for major structural usages. Maximum weight reduction is usually in the range of 30%. Strength levels may be kept reasonably high, but some loss in stiffness (modulus of elasticity) is inevitable, so deflections become more critical. The principal means for achieving weight reduction is using materials other than stone for the coarse aggregate. Some natural materials may be used for this, but for more typical applications synthetic materials are used. One major use is the concrete fill applied on top of formed sheet steel decking in steel-framed structures.

Super Heavyweight Concrete

For some purposes, it may be desirable to achieve an increase in the concrete density (unit weight). A simple means of achieving this is to use a particularly heavy material for the coarse aggregate. Some of the heaviest natural materials are metal ores, but careful analysis must be made of their potential chemical reactions with the concrete.

Insulating Concrete

Use of superlightweight aggregates, usually natural or "popped" mineral materials, together with deliberately entrapped air (foaming), can produce concrete with densities below 30 lb/ft^3. Compressive strength drops to a few hundred psi, so major structural usage is out of the question, but the material is used for the fill on top of roof decks and, in some situations, to insulate steel framing from fire.

Superstrength Concrete

Through the use of specially selected materials, the addition of water-reducing and density-enhancing admixtures, and very special mixing, handling, and curing, concrete strengths in the range of 20,000 psi can now be achieved. The major use to date has been for the lower structures of very tall concrete buildings. This requires a major effort and considerable expertise and is very expensive but is now sometimes accomplished where it offers significant value. The nature of this material is out of the range of traditional procedures and specifications, so its design control is still being developed.

As materials research intensifies—both commercially and with some nonprofit sponsorship—new materials that provide new potential uses are sure to increase in number. Still, traditional sand-and-gravel concrete remains in wide use for common applications.

2

CONSIDERATIONS FOR THE PRODUCTION OF CONCRETE

This chapter presents various issues that must be considered in achieving the forms of concrete construction discussed in this book. While some construction is developed with factory-produced elements of concrete, most ordinary concrete structures still require a large amount of site work. Recognition of the high cost of site work and doing whatever can be done to reduce its cost are critical factors in good design for concrete structures.

2.1 GENERAL CONCERNS FOR CONCRETE

Because concrete is a mixture in which a paste made of portland cement and water binds together fine and coarse particles of inert materials, known as *aggregates,* it is readily seen that by varying the proportions of the ingredients innumerable combinations are possible. These combinations result in concrete of different qualities. When the cement has hydrated, the plastic mass changes to a material resembling stone. In this period of

33

hardening, called *curing*, three things are required: time, favorable temperatures, and the continued presence of water.

To fulfill requirements, it is essential that the hardened concrete have, above all else, *strength* and *durability*. In order for the concrete in its plastic form to be readily placed in the forms, another essential quality is *workability*. When watertightness is required, concrete must be *dense* and *uniform* in quality. Hence, it is seen that, in determining the various proportions of the mixture, the designer must have in mind the purpose for which the concrete is to be used and the exposure to which it will be subjected. The following factors regulate the quality of the concrete: suitable materials, correct proportions, proper methods of mixing and placement, and adequate protection during curing.

2.2 CONCERNS FOR STRUCTURAL CONCRETE

For the production of major structural elements, a principal concern is the specific value of the compressive strength of the finished concrete. This value is assumed in the design work, striven for in the design specifications and the mixing, handling, curing, and general production of the concrete, and if not achieved, can cause major problems with regard to certifying the safety of the structure. Major efforts are made by various parties— designers, concrete producers, builders, testing labs, and code-enforcing agencies—to ensure the proper quality of the finished concrete.

Besides its compressive strength, the other major concern for structural concrete is its general uniformity and well-dispersed quality within the concrete mass. Two principal concerns in this regard are unintentionally trapped air in the form of bubbles and what is described as *segregation* of the mixed materials. Trapped air can result in a sponge like character, rather than a solid character, for the concrete and can also create localized weak points that can become the starting points for structural failures of the concrete material. Vibration of the wet cast concrete is a common means of reducing the amount of trapped air.

Segregation of the materials in the wet mix can result in a layered effect, with the coarser materials settling to the bottom and the water and cement floating to the top. This is largely a matter of careful handling of the materials during construction. Excessive handling, long vertical dropping of the mix, and too much vibration are common causes of segregation.

In some situations, the surface hardness or the general durability of the exposed surfaces may be important for concrete structures. Loss of mass of the concrete and loss of cover for steel reinforcement can occur if surface concrete is worn or eroded away. Special forming, worked fin-

ishes, or applied treatments of surfaces may be used to achieve an enhanced surface for the concrete.

In general, anything done to ensure the best-quality concrete will benefit its structural usage.

2.3 SITECAST CONCRETE

Most concrete for buildings is cast in place; that is, the wet mix is deposited and formed at the place where the finished concrete is desired.
This is now generally referred to as *sitecast concrete* because the location
is usually at a building site. This is compared to *precast concrete,* which
refers to the process of casting elements and then moving them to the
place they are to be used.

Concrete for sitecast construction is typically brought to the site by
the familiar concrete-transporting mixer trucks, with the large rotating
barrels. The mix is prepared at a central batching plant, where controls of
the materials may be carefully monitored. However, the transporting to
the site, proper mixing in the truck, discharging from the truck and depositing in the forms, and handling for placement, finishing, and curing
are all subject to the level of responsibility and craft exercised by the
people involved. Site conditions in terms of accessibility and weather can
be highly critical to the work, requiring extreme measures in some situations to control all the stages in the production process. This book is not
about construction processes or their management, but some awareness
of the issues and limitations is helpful in developing reasonable designs
for concrete structures.

Forming of Sitecast Concrete

An inherent property of concrete is that it may be made in any shape. The
wet mixture is placed in *forms* constructed of wood, metal, or other suitable material, in which it hardens or sets. The forms must be put together
with high-quality workmanship, holding to close dimensional tolerances.
Formwork should be strong enough to support the weight of the wet concrete and rigid enough to maintain position and shape. In addition, formwork should be tight enough to prevent the seepage of water and should
be constructed in such a way as to permit ready removal.

Lumber and plywood used for forms are usually surfaced on the side
that comes in contact with the concrete and frequently are oiled or otherwise sealed. This fills the pores of the wood, reduces absorption of water
from the concrete mixture, produces smoother concrete surfaces, and
permits the forms to be more easily removed.

Steel forms have the decided advantage of being more substantial if the forms are to be reused. Steel gives smoother surfaces to the concrete, although it is almost impossible to avoid showing the joints. For ribbed floors, metal pans and domes are used extensively, and columns, circular in cross section, are invariably made with metal forms.

Because the formwork for a concrete structure constitutes a considerable item in the cost of the completed structure, particular care should be exercised in its design. It is desirable to maintain a repetition of identical units so that the forms may be removed and reused at other locations with a minimum amount of labor.

There are no exact rules concerning the length of time the forms should remain in place. Obviously, they should not be removed until the concrete is strong enough to support its own weight in addition to any loads that may be placed on it.

Placing and Finishing

As soon as cement is mixed with water, a chemical action begins that eventually results in the hardening of the concrete. The first stage of this is the wet mix, which has the character of a thick, viscous fluid. In a short time, however, an initial hardening (called *set)* occurs, and the fluid nature of the mix fades. Before the initial set occurs, the concrete must be fully placed in the forms—a matter of only a few hours with ordinary mixes. Add up the time for loading the trucks at the batching plant, driving the trucks to the site, emptying the trucks, moving the mixed concrete to its desire location, and handling it to fill the forms, and there is not much time for leisure in the operation.

Despite the haste required, the wet concrete must be carefully handled so as not to cause segregation. This can occur if the concrete sits idly in its wet state, is dropped too far when deposited, or is moved around too much in the forms before settling into place. This is a situation that calls for careful supervision but mostly depends on the skills and care of the workers involved.

Surfaces of the concrete in contact with forms will derive a primary form and finish from the surfaces of the forms. Unformed surfaces (usually the top) may receive various treatments, the simplest being a simple struck surface, produced by smoothing with a board or rough wood tool. This is essentially an unfinished surface, which may be additionally treated during the initial hardening or later.

Selection of forming materials may be made to achieve a certain desired form or finish of the cast concrete. Of particular concern is the surface of forms. The smoothest, cleanest surfaces are achieved with steel, fiberglass, or plastic-coated plywood forms. Special surfaces may be achieved with form-lining elements or by coating the forms with various materials. One special material of the latter type is a retarder, which slows down the hardening of the surface cement-water paste, so that the face of the concrete is partly stripped away when the forms are removed, revealing the embedded aggregate just below the surface. Sand-blasting of the freshly cast surface—possibly in combination with use of a retarder— can also produce a roughened, aggregate-exposed surface.

Many other finishes can be achieved, in the process of mixing, initial forming and finishing, or reworking of the hardened surfaces. For structural members, concern must be given to the degree of surface loss that occurs, which may affect loss of member cross sections and concrete cover for the steel reinforcement.

Other concerns for exposed surfaces include the possible effects of various elements used in the construction, of which some examples are:

Bar Chairs and Spacers.   These are used to hold steel bars above the bottom of the forms. They must sit on the forms, and thus their feet will be at the surface when forms are removed. Plastic elements or steel with plastic coating should be used where exposure may cause rust spotting on surfaces.

Form Ties.   Walls are typically formed with two surfaces that are tied to each other across the void (concrete-filled) space to prevent their bowing out from the hydraulic pressure of the wet concrete. These ties become embedded, and their outer ends are hard to conceal.

Joints in Forms.   Large concrete members—long beams, walls, undersides of slabs—must be formed of many units with joints between units. Concrete exposed to view will show these joints, unless the surfaces are completely reworked by some process.

Except for pavements, much structural concrete occurs in basements or foundations, or is covered by other construction, so that surface finishes are often of little concern, other than that they should be reasonably dimensionally true. Dimensional accuracy and reasonable sturdiness may be the only required attributes of forming, and simple struck surfaces may be adequate for nonformed faces of the concrete.

Handling of Freshly Cast Concrete

Regardless of the degree of care taken in proportioning, mixing, and placing, first-quality concrete can be obtained only with due consideration and when adequate provision is made for curing the freshly cast material. The hardening and strength gain of new concrete results from the chemical reaction between the cement and water. This process continues indefinitely as long as water is present and the temperature is favorable. The initial set ordinarily occurs around two to three hours after the concrete has been mixed. Once the concrete is in the forms, exposed surfaces will be subject to water evaporation, and unless this is prevented or retarded, the concrete will develop a poor surface. One such surface contains *crazing,* which consists of many small surface cracks.

Minor loss of the surface may be of little consequence to massive concrete elements. It is potentially quite harmful, however, to thin slabs or walls, where the surface drying can pull other water from the interior mass, resulting in poor material throughout. For surfaces exposed to view, there may also be an issue of the general finish quality of the surface.

To prevent the loss of moisture during curing, several methods may be employed. When hard enough to walk on, slabs may be covered to retard evaporation. Continuous, fine-mist sprinkling is sometimes used. Climate and weather conditions involving air with very low moisture or wind will aggravate this problem.

An ideal curing temperature is around 60°F. At considerably lower temperatures, the chemical reactions will slow down and be incomplete. Of course, the concrete should not be allowed to freeze until it has gained considerable strength. Higher temperatures are sometimes deliberately induced to speed up hardening and strength gain, although at the cost of some loss in the total eventual strength. In general, if it is uncomfortable for people with light clothing, it is uncomfortable for the concrete.

Weather conditions may present only a temporary problem for concrete that will eventually be enclosed within the building. For concrete that will be continuously exposed, such as that for sidewalks and drives, some consideration may be made in designing the mix to enhance its weather resistance. Provision of controlled air entrainment or fiber reinforcement are two such measures. This, however, does nothing for the vulnerability of the freshly cast material.

2.4 DESIGN AND PRODUCTION CONTROLS

Structural designers ordinarily document their work in the form of a set of written computations. These computations will include a listing of design criteria: codes and standards used, concrete strength and steel type used, design loadings assumed, and so on. The computations are concluded by a listing of the design decision information: required shape and dimensions of concrete elements and the positions, number, and size of reinforcing bars. In most cases, some sketches are used in the computations to indicate the arrangement of reinforcing in member cross sections and the locations of bar cutoffs, extensions, bend points, and so on.

Structural computations are not ordinarily used to transmit information to the builder. For this purpose, it is ordinarily the practice to produce a set of contract documents: working drawings and specifications. Translation of the computations into the construction documents is normally done in the design office. The designer should understand this process and be able to check the final form of the construction documents in order to ensure that the designer's instructions have been properly interpreted.

While the construction documents (if thoroughly executed) completely delineate the finished structure, the builders must usually produce a second set of documents that explain more directly to the work force how to make the concrete forms, provide falsework (the supports required during pouring and curing of the concrete), and fabricate and install the steel bars. Although the correctness of this translation is the responsibility of the builder, the designer should also verify the accuracy to avoid mistakes that will delay the construction.

At various stages of the construction, the adequacy of the work should be verified by the designer. The proper shape, details, and dimensions of forms, and the proper installation of reinforcing should be checked prior to pouring of the concrete. Installation in the forms of inserts for attachment, piping, electrical conduit, blocking for ducts, wiring, and piping chases, and so on should be inspected to ensure that they do not critically reduce the structural capacity of affected members.

2.5 INSPECTION AND TESTING

If the operation is of sufficient magnitude, concrete made of various proportions and with aggregates from the sources proposed for use on the job should be tested before construction of the building is started. In the usual procedure, several combinations are tested before the most economical mix that will produce a concrete of the desired strength, density, and workability is chosen. It is customary to continue testing the concrete during construction, particularly if there are changing weather conditions or if a change is made in the sources from which the aggregates are obtained.

The two most common tests of concrete are the test for water content to determine the degree of workability of the fresh concrete and the compression test on cylinders of cured concrete to establish its strength. The effectiveness of these tests in quality control of concrete production depends on obtaining truly representative samples of fresh concrete and following standard procedures during testing. The American Society for Testing and Materials (ASTM) issues ASTM Standards covering sampling and testing, which are prescribed procedures under the ACI Code.

Slump Test

Although the terms *consistency* and *workability* are not strictly synonymous, they are closely related. Consistency may be loosely defined as the wetness of the concrete mixture; it is an index of the ease with which concrete will flow during placement. A concrete is said to be workable if it is readily placed in the forms for which it is intended; for instance, a concrete of given consistency may be workable in large open forms but not in small forms containing numerous reinforcing bars. With this understanding, the slump test may be considered a measure of the workability of fresh concrete.

The equipment for making a slump test consists of a sheet metal truncated cone 12 in. high with a base diameter of 8 in. and a top diameter of 4 in. Both top and bottom are open. Handles are attached to the outside of the mold. When a test is made, freshly mixed concrete is placed in the mold in a stipulated number of layers and each is rodded separately a specified number of times with a steel rod. When the mold is filled to the top, it is lifted off at once. The slump of the molded concrete is measured by taking the difference in height between the top of the mold and the top of the slumped mass of concrete (Figure 2.1).

If the concrete settles 3 in., we say that the particular sample has a 3-in. slump. Thus, the degree of consistency of the concrete is ascertained. Recommended slump ranges for various types of construction are given in construction standards.

Compression Test

Tests of compressive strength are made at periods 7 and 28 days on specimens prepared and cured in accordance with prescribed ASTM testing procedures. The specimen to be tested is cylindrical in shape and has a height twice its diameter. The standard cylinder is 6 in. in diameter and 12 in. high when the maximum size of the coarse aggregate does not exceed 2 in. For larger aggregates, the cylinder should have a diameter at least three times the maximum size of the aggregate, and its height should be twice the diameter.

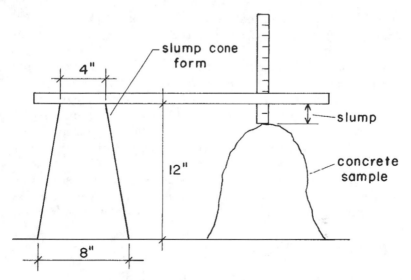

Figure 2.1   Slump test for wet concrete.

The mold used for the cylinders is made from metal or other nonabsorbent material, such as paraffin-coated cardboard. It is placed on a smooth plane surface (glass or metal plate) and filled with freshly made concrete. As soon as casting of the cylinder is complete, the top of the specimen is covered to prevent the concrete from drying.

At the end of the curing period, specimens are placed in a testing machine and a gradually increasing compressive load is applied until the specimen fails. The load causing failure is recorded, and this load, divided by the cross-sectional area of the cylinder, gives the ultimate compressive unit stress of the specimen. The same test is made on other specimens taken at the same time and cured under similar conditions, which, of course, results in a range of values for compressive strength.

Depending on structural application, climate conditions, or other factors, other tests may be necessary for any given project.

2.6 INSTALLATION OF REINFORCEMENT

To facilitate the shop fabrication and field installation of reinforcing bars, the bar supplier usually prepares a set of drawings—commonly called *shop drawings.* These drawings consist of the supplier's interpretation of the engineering contract drawings, with the information necessary for the workers who fabricate the bars in the shop and those who install the bars in the field prior to pouring of the concrete. The exact cut lengths of bars, the location of all bends, the number of each type of bar, and so forth will be indicated on these drawings. While the correctness of these drawings is the responsibility of the supplier, it is usually a good idea for the designer to verify the drawings in order to reduce mistakes in the construction.

Reinforcing bars must be held firmly in place during the pouring of the concrete. Horizontal bars must be held up above the forms; vertical bars must be braced from swaying against the forms. The positioning and holding of bars is done through the use of various accessories and a lot of light-gauge tie wire. When concrete surfaces are to be exposed to view after being poured, it behooves the designer to be aware of the various problems of holding bars and bracing forms since many of the accessories used ordinarily will be partly in view on the surface of the finished concrete.

Installation of reinforcing may be relatively simple and easy to achieve, as in the case of a simple footing or a single beam. In other cases, where the reinforcing is extensive or complex, the problems of installation may require consideration during the design of the members. When beams intersect each other, or when beams intersect columns, the extended bars from the separate members must pass each other at the joint. Consideration of the "traffic" of the intersecting bars at such joints may affect the positioning of bars in the individual members.

2.7 PRECAST CONCRETE

Precasting refers to the process of construction in which a concrete element is cast somewhere other than where it is to be used. The other place may be somewhere else on the building site or away from the site, probably in a casting yard or factory. The precast element may be prestressed, may be of ordinary reinforced construction, or may even be without reinforcement. The single precast element may be a component of a general precast concrete system or may serve a singular purpose in a construction system of mixed materials or types of elements. This section gives some discussion of the uses of precasting and the problems encountered in designing precast elements and systems.

Use of Precasting

The technique of precasting is utilized in a variety of ways. Undoubtedly, the most widely used precast element is the ordinary concrete block—

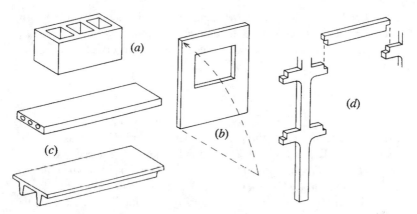

Figure 2.2 Structural elements of precast concrete: (*a*) CMU, (*b*) tilt-up wall panel, (*c*) hollow-core deck unit and double-T deck unit, and (*d*) column and beam frame.

called CMU (concrete masonry unit). Most structural masonry is made from these units. In ordinary construction, the block form shown in Figure 2.2*a* is commonly used; for reinforced masonry construction (used exclusively in zones of high seismic risk), a different form is used. Construction with CMUs is discussed in Section 2.9.

Another widely used precast element is the tilt-up wall unit, shown in Figure 2.2*b*. This element is sitecast in a horizontal position, then tilted up and moved by a crane to its desired location. The casting bed usually consists of the building floor slab on grade, resulting in a considerable reduction in forming cost. This type of construction is widely used for one-story and low-rise commercial structures in the southern and western regions of the United States.

As discussed in Section 1.7, some of the most widely used prestressed elements are the flat-spanning units of hollow core or tee form used for roof and floor construction. These units are produced in casting factories in continuous production processes.

Structural systems consisting of connected components of precast concrete have been produced in great variety. Some of these have been produced as patented, manufactured systems, but mostly they have been the single, innovative products of individual designers.

Design and construction of precast concrete is strongly influenced by the standards and publications of the Precast Concrete Institute (PCI). Industry products are largely developed in conformance with these standards. Anyone contemplating the design of a unique system or element of precast concrete is advised to investigate the information available from PCI.

Advantages of Precast Concrete

There are various reasons for considering the use of precast concrete construction. In some cases, the choice is between precasting and ordinary construction of cast-in-place concrete, with elements formed and cast at the location where they are to be used. In other cases, the choice may be between using precast elements or some other material or type of construction. The following are some advantages offered by the precasting process, generally in comparison to cast-in-place construction:

Faster Site Work.   Cast-in-place concrete construction usually proceeds quite slowly, requiring construction of forms, installation of reinforcing, pouring of concrete, hardening to sufficient strength to permit removal of forms, and so on. Erection of precast elements is more akin to construction with steel or timber structures, and faster site work may be an advantage where construction time is highly constrained. However, it is the total building construction time that is significant, not just the time to get up the structure. If time cannot also be gained in other parallel construction activities, the rapidly erected structure may just sit there and wait for the other project work to catch up.

Forming Economies.   For the ordinary cast-in-place concrete structure, a major portion of the total cost is represented by the forming. This includes the cost of the construction, bracing and support, and removal of the forming. Some reuse of items may be possible, but the process tends to use materials up rapidly. Precasting offers more potential for reuse of forms, even with sitecast construction. Factory processes involve extensive reuse of forms or production by forming processes such as extrusion. Reduction of on-site labor costs may often be the major gain in this area.

Quality Control.   Precision of detail, quality, and uniformity of finishes, and uniformity of concrete properties (color, density, compaction, etc.) may be ensured in factory production to a degree not possible with sitecasting. Here it is not just precasting but factory conditions that are the issue. All of this is even truer if the element is a standard manufactured product subject to ongoing quality control in its production. This is, obviously, of greatest concern for construction elements exposed to view, especially wall components.

Use of Predesigned Elements and Systems. Face it—design effort means time and money. Use of a standard predesigned building construction component means a shortcut in design development effort. If the component is part of a system, with system-wide concerns for total building utilization carefully considered and standardized, the savings in design effort and the relative assurance of end results may be quite significant.

Utilization of High-Quality Concrete and Prestressing. The potential of utilization of structural properties of very-high-strength concrete usually occurs in association with prestressing (versus ordinary reinforcing). Factory-produced concrete is routinely of higher quality than the concrete produced by sitecasting. In some situations, denser surfaces, lower permeability, and reduced shrinkage may also be significant. This may be a factor in the use of mixed systems of precast and cast-in-place elements.

Problems with Precast Concrete

As with any form of construction, there are some particular problems associated with precast concrete. These are not necessarily insurmountable, but designers should be aware of them and of the considerations that may be required in using the construction method. The following are some major concerns that may be of significance in various situations:

Handling and Transporting. Concrete construction is usually heavy—precast elements included. Precast units are often of considerable size, and the combination presents a major problem of handling and transporting the heavy and relatively fragile units. Stresses induced during handling and erecting units may be significant structural design concerns. Use of factory-cast units is usually feasible only within some reasonable distance from the factory.

Cost. The cost of production, handling, and transporting of precast units is considerable. There needs to be some considerable list of other advantages to make this form of construction generally competitive—not so much with alternative cast-in-place concrete but with other materials and systems. In the end, the total building construction cost is most significant, not just structural cost.

Connections.   From a general construction development point of view, the single biggest problem in design with precast elements is usually the connection of elements. From a structural response consideration, it is here that the major difference occurs between precast and ordinary cast-in-place concrete construction. The completely precast structure has more in common with structures of steel and timber than with ordinary concrete structures. The adequate development of individual connections is a problem, but almost more significant is the overall loss of natural continuity and inherent stability of the fully cast-in-place concrete construction. A major effort by the PCI is in the development of recommended practices for design and construction of connections of precast elements.

Integration.   This refers to the problem of incorporating the precast concrete elements into the general building construction. A major problem to be dealt with in this regard is the loss of some opportunities that are present with other types of construction. Installation of hidden items, such as wiring, piping, ducts, and housing for light switches, power outlets, recessed lighting fixtures, bathroom medicine cabinets, fire hose cabinets, and exit signs, is made somewhat more difficult. With light wood frames, ordinary cast-in-place concrete, and most other types of construction, precise locating of these items and the actual installation can be done during site construction work. With precast concrete construction, provisions must be made in advance—a procedure that is not impossible, but that does not fit with the routine operations with which most designers and builders are familiar. This is really a minor matter of adjustment of procedures and scheduling but one that can create difficulties if it is not anticipated and provided for.

Seismic Effects.   Usage of precast concrete has received some setbacks in response to the performance of some structures during major seismic events. Promotional literature of the wood and steel industry frequently contains some dramatic pictures of collapsed structures of precast concrete. This is indeed a major problem to be dealt with in design, especially for designing connections of structural components. The loss of continuity in changing from cast-in-place to precast construction is a major concern. Learning from failures (a basic process in all structural design areas), the precast industry has developed more stringent criteria and techniques. Nevertheless, the seismic response of heavy, individually stiff and brittle elements, with many joints in the assembled system, makes the precast structural system less than ideal in seismic response. In spite of this, some systems—such as the tilt-up wall—see ongoing extensive use in regions of high seismic risk, attesting to the fact that design can be effectively achieved when other factors are sufficiently persuasive in the overall decision of system selection.

2.8 MIXED SYSTEMS: SITECAST AND PRECAST

The completely precast concrete structure does not represent the widest usage of precasting. Precast components are used in many situations in conjunction with a building structure and general construction with other materials and systems. This is, of course, the *usual* situation; few buildings are all steel, all wood, all concrete, or all masonry.

Precast concrete decks, especially of the hollow-cored form, are used with frames of steel or sitecast concrete and with bearing walls of masonry or sitecast concrete. Precast concrete wall panels are used with structures of steel, wood, and sitecast concrete. The separate component/separate material situation is a common one in building construction. Of course, the individual functions of the components and the interfacing of components for structural interaction must be dealt with in design and in the development of proper construction details.

The blending of components of precast and sitecast concrete offers some opportunities for complementary enhancement of the two methods. Figure 2.3a shows a joint detail commonly used with tilt-up wall panels. In this case, a formed and site-poured concrete column is used to effect the structural connection between two panels, as well as to achieve a positive interaction of the panels (as shear walls) with the continuous poured concrete frame. A similar situation is shown in Figure 2.3b, in which two levels of a structure, consisting of slab and beam concrete systems, are connected to a precast concrete wall panel.

A slightly different situation is shown in Figure 2.3c. In this case, precast concrete units are used as forms for sitecast concrete columns and the edges of a sitecast framing system. Here, the precast units do not serve significant structural tasks in their own right, in comparison to the situations in Figures 2.3a and b. They do, however, eliminate the need for other, temporary forming. More importantly, they provide the possibility

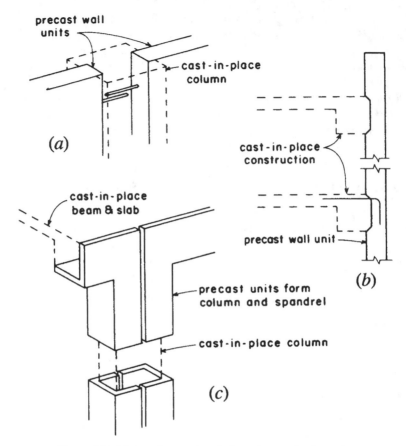

Figure 2.3   Mixing of precast and sitecast concrete elements.

for a quality of detail and textural control of the exposed concrete that cannot easily be achieved with sitecast concrete.

2.9   CONCRETE MASONRY

In times past, structural masonry was achieved primarily with stone and bricks. Concrete, in a crude form, was used as filler to allow large elements to be constructed with shells of finer material. At present, in the United States, much of what has the appearance of a masonry structure is really achieved with a thin veneer over some other construction—frequently a wood or steel frame. Structural masonry, when it actually occurs, is sometimes of brick, but most often uses masonry units of precast concrete.

Structural masonry is used most often for walls, and a discussion of wall construction is given in Section 12.6. Use of concrete masonry walls is also illustrated in the building system design examples in Chapter 16. Columns made from CMUs are discussed in Chapter 10.

3

GENERAL REQUIREMENTS FOR REINFORCED CONCRETE STRUCTURES

Design and production of structures of reinforced concrete are controlled by many regulations, industry standards, and practical considerations. This chapter presents some of the major considerations that apply to all concrete structures. Detailed requirements for various types of concrete structural elements are discussed in the chapters that deal with the general design of those elements. Some broader considerations for the development of building structural systems are discussed in Chapters 15 and 16.

3.1   CODE AND INDUSTRY STANDARDS

Standards of practice for structural design and general construction evolve slowly from shared experiences and research. New theories and innovative designs or construction techniques are tested and eventually accepted or rejected. Materials and processes for production and construction are steadily improved. At any point in time, the distilled residue and essence of this are reflected in the latest "standards of practice," presented as building codes and industry standards. Some of the major sources for these standards are discussed in the Introduction.

General practices in design and construction work must acknowledge these basic reference sources. However, published standards are often minimal in nature, specifying the least that must be done, not the best that can be done. Where something more than minimal results is to be expected, satisfying minimal requirements is not likely to be the ideal objective.

Operative standards of practice influence various activities with regard to the design and construction of concrete structures. Major areas of consideration are:

- Design methods and criteria
- Production and construction processes
- Required tests and certifications
- Code requirements for fire resistance
- General code requirements that affect planning and detailing of building construction

As mentioned previously, these matters are discussed in detail in many other parts of this book

3.2 PRACTICAL CONSIDERATIONS

Producing concrete structures is a real-world situation with many very pragmatic concerns. Designers, generally not directly involved in the production work, must relate practically to these concerns or suffer many embarrassments when their speculations face reality. The following are some basic, practical concerns for concrete construction:

Maximum Single Pour

Various practical limits ordinarily establish the maximum amount of concrete that can be placed at one time. For large structures, this is usually only a fraction of the entire structure. Thus, the so-called *cold joint*, or *construction joint*, occurs when pouring stops for some significant time. The concrete poured hardens, and when pouring continues, the new, wet concrete is poured against the old, hardened concrete. These joints must be anticipated and preferably incorporated into the design and detailing of the construction.

The size of pour may be limited by time (the 8-hour workday), by the size of the work crew, by the accessibility of the site, by the number of concrete trucks available for delivery, by the method for placing concrete, or by practical limits of the form of construction. An example of the latter is a multistory structure, where a practical pour limit is one story at a time.

For sitecast structures, the entire structure is typically considered as a continuous, monolithic mass. Achieving this in fact requires some careful considerations for the effects of the construction joints.

Concrete Design Strength (f'_c)

For structural design, a major early design decision is the design strength of the concrete. This must be related fundamentally to the nature of the structure, but also to practical considerations of the capabilities of the current technology, the abilities of the available workers and contractors, and the size and budget of the project. Thus, some designs may push against the limits of the current technology to achieve the best possible concrete (for example, for major high-rise structures), while others may best use a threshold, minimum level of material. This is a very critical issue but quite difficult to generalize simply.

Accuracy of Construction

Sitecast concrete is very rough work, not generally subject to precise, refined detail and finish. Experience with actual construction will bring some judgment about what must be tolerated and what can be improved with careful specifications, choices for forming materials, and some extra efforts at site inspection of the work.

In general, off-site, plant-cast concrete is capable of achieving higher quality in terms of the concrete material itself, as well as dimensional accuracies and finishes of the cast elements. This may not be so critical to the production of basic structural components but allows for production of truly superior elements for cladding and other architectural finishing of the construction.

Of course, if the concrete work is to be covered with applied finishes or generally encased in the finished construction, its natural rough character may be of little consequence. Consideration should be given, however, to the tolerances required for attachments of more refined elements of the construction, recognizing the limits of accuracy of the concrete structure.

Minimum Size of Concrete Members

For practical reasons, as well as the satisfying of various requirements for cover and bar spacing, there are minimum usable dimensions for various reinforced concrete members. When flexural reinforcement is required in slabs, walls, or beams, its effectiveness will be determined in part by the distance between the tension-carrying steel and the far edge of the compression carrying concrete. Thus, extremely shallow beams and thin slabs or walls will have reduced efficiency for flexure.

In slabs and walls, it is usually necessary to provide two-way reinforcing. Even where the bending actions occur in only one direction, the code requires a minimum amount of reinforcing in the other direction for control of cracking due to shrinkage and temperature changes. As shown in Figure 3.1a, even with minimum cover and small bars, a minimum slab thickness is approximately 2 in. Except for joist or waffle construction, however, slab thicknesses are usually greater, for reasons of development of practical levels of flexural resistance. Thus, reinforcing is more often as shown in Figure 3.1b, with the bars closer to the top or bottom, depending on whether the moment is positive or negative.

In many cases, it is desirable for the slab to have a significant fire rating. Building codes often require additional cover for this purpose, and typically specify minimum slab thicknesses of 4 in. or more. Slab thicknesses required for this purpose also depend on the type of aggregate that is used for the concrete.

Walls of 10 in. or greater thickness often have two separate layers of reinforcing, as shown in Figure 3.1c. Each layer is placed as close as the requirements for cover permit to the outside wall surface. Walls with crisscrossed reinforcing (both vertical and horizontal bars) are seldom made less than 6 in. thick.

As shown in Figure 3.1d, concrete beams usually have a minimum of two reinforcing bars and a stirrup or tie of at least No. 2 or No. 3 size. Even with small bars, the minimum beam width in this situation is at least 8 in., with 10 in. being much more practical.

For rectangular columns with ties, a limit of 8 in. is usual for one side of an oblong cross section and 10 in. for a square section. Round columns may be either tied or spiral wrapped. A 10-in. diameter may be possible for a round tied column, but 12 in. is more practical and is the usual mini-

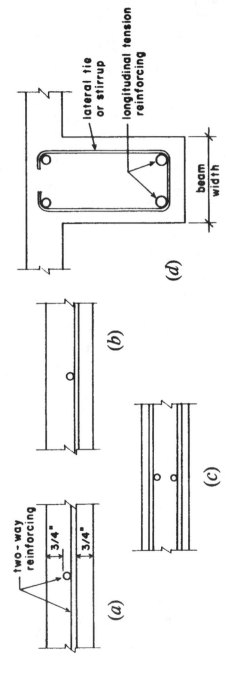

Figure 3.1 Placement of steel reinforcement in concrete structural elements.

mum for a spiral column, with larger sizes required where more cover is necessary.

3.3   CONTROL OF CRACKING

All rigid, tensile-weak materials are highly susceptible to some form of cracking. This is a major concern for concrete, as well as for stone, masonry, and plaster in building construction uses.

Sources of Cracking

The major sources of cracking in concrete are the following:

Shrinkage.   The water-mixed material shrinks in volume when it dries. In ordinary concrete, this is typically about a 2 to 3% volume change. Long elements, thin elements, and continuous elements with many geometric discontinuities are most susceptible to shrinkage cracking. As concrete building structures ordinarily have all of these, some design considerations must usually be made.

Temperature Change.   Thermal expansion and contraction is similar to shrinkage in terms of the effects of volume change. This is of particular concern in regions with a very wide range of outdoor temperatures and applies especially to structures exposed to the weather.

Structural Action.   Any structural action other than simple direct compression will produce some tensile stress in the material of a structure. Where major bending or torsion occurs, this is usually the most severe. In ordinary reinforced concrete beams and spanning slabs, some cracking of the concrete in the vicinity of the tensile reinforcement is inevitable. (See discussion in Section 6.1.)

Settlement of Supports.   Concrete structures are usually quite heavy and are consequently vulnerable to settlement on soils in which this offers a problem. Add this to the usual rigid nature of concrete structures and the potential for problems is considerable.

Means for Crack Control

Cracking of concrete structures is a major concern. Control of cracking is seldom achieved by any single means, but typically by some combination of the following:

Minimum Reinforcement.   Cracks are essentially tension stress failures, and a basic crack-resisting technique is the use of steel reinforcement. Use of minimum reinforcement for resistance to the effects of shrinkage and thermal change is discussed in Section 3.4.

Control Joints.   A major means for crack control is the establishment of joints that interrupt the continuity of the concrete mass, in effect functioning as preestablished cracks. Specific considerations for these are discussed in Section 14.1 (for paving slabs) and in Section 12.1 (for walls). Locations of control joints should preferably be developed with logical consideration for practical construction joints. Structural design may also indicate the need for control joints for thermal expansion of the general construction, seismic separation, control of undesirable continuity effects, or other structural purposes; these joints will also function for crack control. For good design, any visible joints should be expressed in the general architectural design and detailing of the building.

Fiber Reinforcement.   Although currently used mostly for nonstructural elements of the construction (roofing tiles, cladding panels, etc.), the inclusion of fiber materials in the concrete mix enhances the general tensile resistance of the basic concrete material, resulting in some general reduction of minor cracking.

Prestressing.   As discussed in Section 1.7, the use of prestressing is a means for elimination of cracking caused by flexure in beams and other structural components. This is seldom the principal reason for using prestressing but is nevertheless an added positive factor.

3.4   GENERAL REQUIREMENTS FOR STEEL REINFORCEMENT

The following are some general requirements for steel reinforcement in reinforced concrete structures. Specific requirements for individual types of structural components are also presented in other parts of this book.

Minimum Reinforcement.   In the design of most reinforced concrete members, the amount of steel reinforcing required is determined from computations and represents the amount determined to be necessary to resist the required tensile force in the member. In various situations, however, there is a minimum amount of reinforcing that is desirable, which may on occasion exceed that determined by the computations. The ACI

Code makes provisions for such minimum reinforcing in columns, beams, slabs, and walls. The minimum reinforcing may be specified as a minimum percentage of the member cross-sectional area, as a minimum number of bars, or as a minimum bar size. These requirements are discussed in the sections that deal with the design of the various types of members.

Shrinkage and Temperature Reinforcement. The essential purpose of steel reinforcing is to prevent the cracking of the concrete due to tension stresses. In the design of concrete structures, investigation is made for the anticipated structural actions that will produce tensile stress: primarily the actions of bending, shear, and torsion. However, tension can also be induced by the shrinkage of the concrete during its drying out after the initial pour. Temperature variations may also induce tension in various situations. To provide for these later actions, the ACI Code requires a minimum amount of reinforcing in such members as walls and slabs even when structural actions do not indicate any need. These requirements are discussed in the sections that deal with the design of these members.

Cover. Cover of the steel bars must be provided for various reasons. A prime concern is simply the need for the concrete to "grab" the steel by surrounding it, so that the two materials can truly interact in the response of the composite structure to needed structural actions. Beyond this are the practical concerns for weather protection, general protection from air and moisture that causes rusting, and insulation for fire protection of the steel. Some code limits for cover are discussed in Section 1.6. Specific concerns for particular types of structural components are discussed in other chapters, and some detailed concerns are presented in the design examples in Chapter 16.

The outer dimensions of the concrete in a given structural member combine with the required cover dimensions to define a limiting space for the placing of the reinforcement. Within this space, the necessary separation of the individual bars—as discussed next—determines how much reinforcement can actually be placed inside the defined concrete member. If the space provided is not adequate for the amount of steel required, redesign of the dimensions of the concrete member must usually be considered.

Spacing of Steel Bars. Where multiple bars are used in members (which is the common situation), there are both upper and lower limits for the spacing of the bars. Lower limits are intended to permit adequate

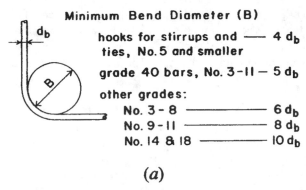

(a)

Figure 3.2 Bend requirements for steel reinforcing bars.

development of the concrete-to-steel stress transfers and to facilitate the flow of the wet concrete during pouring. For columns, the minimum clear distance between bars is specified as 1.5 times the bar diameter or a minimum of 1.5 in. For other situations, the minimum is one bar diameter or a minimum of 1 in.

For walls and slabs, maximum center-to-center bar spacing is specified as three times the wall or slab thickness or a maximum of 18 in. This applies to reinforcement required for computed stresses. For reinforcement that is required for control of cracking due to shrinkage or temperature change, the maximum spacing is five times the wall or slab thickness or a maximum of 18 in.

For adequate placement of the concrete, the largest size of coarse aggregate should be not greater than three-fourths of the clear distance between bars.

Bending of Reinforcement.   In various situations, it is sometimes necessary to bend reinforcing bars. Bending should be done in the fabricating shop instead of at the job site, and the bend diameter (see Figure 3.2a) should be adequate to avoid cracking the bar. Bending of bars is sometimes done in order to provide anchorage for the bars. The code defines such a bend as a "standard hook," and the requirements for the details of this type of bend are given in Figure 8.2. As the yield stress of the steel is raised, bending—which involves yield stress development—becomes increasingly difficult. Bending of bars should be avoided when the yield stress exceeds 60 ksi [414 MPa], and where it is necessary, should be done with bend diameters slightly greater than those given in Figure 3.2.

4

INVESTIGATION AND DESIGN OF REINFORCED CONCRETE

Investigation of structures, whether pursued simply for the process of acquiring general knowledge or useful data or pursued for a more intelligent solution of some specific problem, is itself a major field of study. The material in this chapter consists of discussions of the nature, purposes, and various techniques of the investigation of structures. As in all of the work in this book, the primary focus is on material relevant to the tasks of structural design.

4.1 SITUATIONS FOR INVESTIGATION AND DESIGN

Most structures exist to perform a task. Their evaluation must, therefore, begin with consideration of the effectiveness with which they satisfy the usage requirements determined by that task. Three elements of this effectiveness should be considered: the structure's functionality, feasibility, and safety.

Functionality refers to the various attributes of the structure, such as its shape, detail, durability, fire resistance, and weather resistance, as these relate to its use. *Feasibility* includes consideration of the dollar cost, the availability of materials, and the overall practicality of achieving its construction. *Safety,* in terms of structural actions, is generally obtained by having some margin between the structure's capacity for resistance and the demands placed on it. Investigation of a structure, as related to the effectiveness of design work, must deal with all of these concerns.

The purpose of investigation for safety, consisting of an analysis of a structure's actions under some loading condition, is to establish an understanding of its behavior. By this means, its working responses to service demand conditions can be understood and evaluated. If the loading condition is extended to the point of failure of the structure, the ultimate limit of the structure's capability can be quantified. It is useful for various design purposes to understand both of these: the actions under service load conditions and the ultimate resistance of the structure.

Analysis for investigation may progress with the following considerations:

1. Determination of the structure's physical nature with regard to material, form, scale, detail, orientation, location, support conditions, and internal character
2. Determination of the demands placed on the structure; that is, the loads and the manner of their application, and any usage limits on the structure's deformation
3. Determination of the nature of the structure's responses to the demands (loads), in terms of support forces, deformations, and internal stresses
4. Determination of the limits of the structure's capabilities, including its ultimate (maximum) load resistance
5. General evaluation of the structure's effectiveness

Analysis can be performed in several ways. The nature of the structure's deformation under load can be visualized, with mental images or sketches. Using available theories and techniques, mathematical models of the structure can be created and manipulated. Finally, the structure itself can be built and tested under a simulated load, or a scaled model can be built and tested in a laboratory.

When reasonably precise quantitative evaluations are required, the most useful tools are direct measurements of physical responses and careful mathematical modeling with theories and procedures that have been demonstrated to be reliable. Ordinarily, mathematical modeling of some kind must precede actual construction, even that of a test model. Direct measurement is usually limited to experimental studies or to efforts to verify questionable theories or techniques.

Use of physical testing in the design process for building structures is rare. Most design work is based on experience, with some support from computations using commonly accepted procedures and data. However, in order to support the procedures for mathematical analysis and provide data for computations, a great amount of testing has been done on laboratory specimens of reinforced concrete. Thus, while individual design projects seldom utilize physical testing, the design procedures and data used have been developed with extensive verification from physical tests.

4.2 METHODS OF INVESTIGATION AND DESIGN

Traditional structural design was developed primarily with a method now referred to as *stress design*. This method utilizes basic relationships derived from classic theories of elastic behavior of materials, and the adequacy or safety of designs are measured by comparison with two primary limits: an acceptable level for maximum stress and a tolerable limit for the extent of deformation (deflection, stretch, etc.). These limits are calculated as they occur in response to the service loads; that is, the loads caused by the normal usage conditions visualized for the structure. This method is also called the *working stress method,* the stress limits are called *allowable working stresses,* and the tolerable movements are called *allowable deflection, allowable elongation,* and so on.

In order to convincingly establish both stress and strain limits, it was necessary to perform tests on actual structures. This was done extensively, both in the field (on real structures) and in testing laboratories (on specimen prototypes or models). When nature provides its own tests in the form of structural failures, forensic studies are typically made extensively by various people—for research or establishment of liability.

Testing has helped to prove, or disprove, the design theories and to provide data for the shaping of the processes into an intelligent operation. The limits for stress levels and for magnitudes of deformation—essential to the working stress method—have been established in this manner. Thus, although a difference is clearly seen between the stress and strength methods, they are actually both based on evaluations of the total capacity of structures tested to their failure limits. The difference is not minor, but it is really mostly one of procedure.

4.3   THE STRESS METHOD

The stress method generally consists of the following:

1. The service load (working load) conditions are visualized and quantified as intelligently as possible. Adjustments may be made here by the determination of various statistically likely load combinations (dead load plus live load plus wind load, etc.), by consideration of load duration, and so on.
2. Stress, stability, and deformation limits are set by standards for the various responses of the structure to the loads: in tension, bending, shear, buckling, deflection, and so on.
3. The structure is then evaluated (investigated) for its adequacy or is proposed (designed) for an adequate response.

An advantage obtained in working with the stress method is that the real usage condition (or at least an intelligent guess about it) is kept continuously in mind. The principal disadvantage comes from its detached nature regarding real failure conditions because most structures develop much different forms of stress and strain as they approach their failure limits.

4.4   THE STRENGTH METHOD

In essence, the working stress method consists of designing a structure to work at some established appropriate percentage of its total capacity. The strength method consists of designing a structure to fail, but at a load condition well beyond what it should have to experience in use. A major reason for the favoring of strength methods is that the failure of a structure is relatively easily demonstrated by physical testing. What is truly appropriate as a working condition, however, is pretty much a theoretical speculation. In any event, the strength method is now largely preferred in professional design work. It was first mostly developed for design of reinforced concrete structures but is now generally taking over all areas of structural design work.

Nevertheless, it is considered necessary to study the classic theories of elastic behavior as a basis for visualization of the general ways that structures work. Ultimate responses are usually some form of variant from the classic responses (because of inelastic materials, secondary effects, multimode responses, etc.). In other words, the usual study procedure is to first consider a classic, elastic response, and then to observe (or speculate about) what happens as failure limits are approached.

For the strength method, the process is:

1. The service loads are quantified as in step 1 for the stress method and then are multiplied by an adjustment factor (essentially a safety factor) to produce the factored load.

2. The form of response of the structure is visualized and its ultimate (maximum, failure) resistance is quantified in appropriate terms (resistance to compression, to buckling, to bending, etc.). Sometimes this quantified resistance is also subject to an adjustment factor called the *resistance factor*.

3. The usable resistance of the structure is then compared to the ultimate resistance required (an investigation procedure), or a structure with an appropriate resistance is proposed (a design procedure).

When the design process using the strength method employs both load and resistance factors, it is now called *load and resistance factor design* (*LRED*). The procedures used in this method are presented in Chapter 5.

4.5 INVESTIGATION OF COLUMNS AND BEAMS

Structural investigation begins with an overall analysis of the entire structure to determine responses at supports (reactions) and the type and magnitudes of interior force actions (tension, compression, shear, bending, and torsion). For systems composed of simple beams and individual, pin-ended columns (as in most wood frames, for example), this analysis is quite easily performed. Most concrete frame structures, on the other hand, have members that are continuous through many spans and beam and column groups that constitute rigid frames. Concrete frames are, thus, commonly quite statically indeterminate, and their investigation is complex. Analysis of complex indeterminate structures is beyond the scope of this book, but the following materials in this chapter are provided to explain some aspects of their behavior. For approximate design, various approximation methods may be used, as discussed in the last section of this chapter.

Investigation of Columns

Elementary considerations for columns begin with basic concern for direct, axial compression, with possible concern for any potential for buckling (failure due partly to slenderness of the column). In rigid frame structures, however, columns are also subjected to bending and shear. In the case of the three-dimensional rigid frame—the common situation for building frame structures—there is typically bending in two directions and sometimes additional torsional twisting.

For concrete columns, there is also concern for the feasibility of precise construction, so that even for simple, axially loaded columns, some bending is required to be assumed. The general concerns for analysis and design of columns are discussed in Chapter 10. Some aspects of column-beam frame behavior are discussed in this chapter and more extensively in Chapter 11.

Investigation of Beams

The simple, single-span beam is a rare situation in reinforced concrete structures. As shown in Figure 4.1, the simple beam may exist when a single span is supported on bearing-type supports that offer little restraint (Figure 4.1a), or when beams are connected to columns with connections that offer little moment resistance (Figure 4.1b). Although these situations are common in structures of steel and wood, they seldom occur in concrete structures, except when precast elements are used. For single-story structures, supported on bearing-type supports, continuity resulting in complex bending can occur when the spanning members are extended

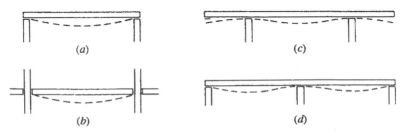

(a) (c)

(b) (d)

Figure 4.1   Flexural deformation of various beam forms.

over the supports. This may occur in the form of cantilevered ends (Figure 4.1c) or of multiple spans (Figure 4.1d). These conditions are common in wood and steel structures, and can also occur in reinforced concrete structures. Members of steel and wood are usually constant in cross section throughout their length; thus, it is necessary only to find the single maximum value for shear and the single maximum value for moment. For the concrete member, however, the variations of shear and moment along the beam length must be considered, and several different cross sections must be investigated.

Figure 4.2 shows conditions that are common in concrete structures when beams and columns are cast monolithically. For the single-story structure (Figure 4.2a), the rigid joint between the beam and its supporting columns will result in behavior shown in Figure 4.2b; with the columns offering some degree of restraint to the rotation of the beam ends. Thus some moment will be added to the tops of the columns and the beam will behave as for the center portion of the span in Figure 4.1c, with both positive and negative moments. For the multistory, multispan concrete frame, the typical behavior will be as shown in Figures 4.2c and 4.2d. The columns above and below, plus the beams in adjacent spans, will contribute to the development of restraint for the ends of an individual beam span. This condition occurs in steel structures only when welded or heavily bolted moment-resisting connections are used. In sitecast concrete structures, it is the normal condition.

The structures shown in Figures 4.1d, 4.2a, and 4.2c are statically indeterminate. This means that their investigation cannot be performed using only the conditions of static equilibrium. Although a complete considera-

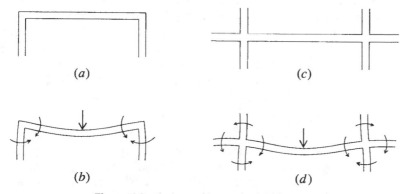

Figure 4.2 Actions of beams in rigid frames.

tion of statically indeterminate behaviors is well beyond the scope of this book, some treatment must be given for a realistic development of the topic of design of reinforced concrete structures. The discussions that follow will serve to illustrate the various factors in the behavior of continuous frames and will provide material for approximate analysis of common situations.

Effects of Beam End Restraint

Figures 4.3*a* to *d* shows the effects of various end support conditions on a single-span beam with a uniformly distributed load. Similarly, Figures 4.3*e* to *h* shows the conditions for a beam with a single concentrated load. Values are indicated for the maximum shears and moments for each case. (Values for end reaction forces are not indicated since they are the same as the end shears.)

We note the following for the four cases of end support conditions:

1. Figures 4.3*a* and *e* show the cantilever beam, supported at only one end with a *fixed-end* condition. Both shear and moment are critical at the fixed end, and maximum deflection occurs at the unsupported end.

2. Figures 4.3*b* and *f* show the classic "simple" beam, with supports offering only vertical force resistance. We will refer to this type of support as a *free end* (meaning that it is free of rotational restraint). Shear is critical at the supports, and both moment and deflection are maximum at the center of the span.

3. Figures 4.3*c* and *g* show a beam with one free end and one fixed end. This support condition produces an unsymmetrical situation for the vertical reactions and the shear. The critical shear occurs at the fixed end, but both ends must be investigated separately for the concrete beam. Both positive and negative moments occur, with the maximum moment being the negative one at the fixed end. Maximum deflection will occur at some point slightly closer to the free end.

4. Figures 4.3*d* and *h* show the beam with both ends fixed. This symmetrical support condition results in a symmetrical situation for the reactions, shear, and moments with the maximum deflection occurring at midspan. It may be noted that the shear diagram is the same as for the simple beams in Figures 4.3*b* and *f.*

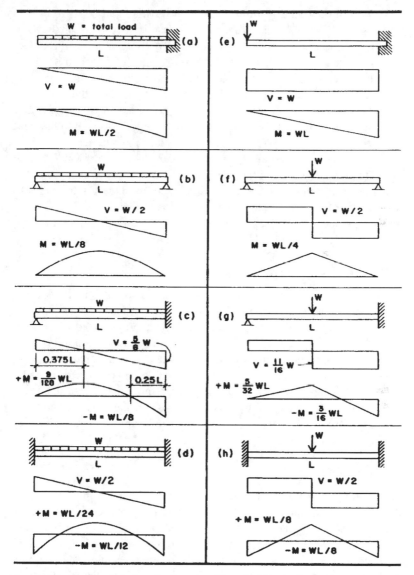

Figure 4.3 Response values for beams with uniformly distributed loading and single, concentrated loads.

Continuity and end restraint have both positive and negative effects with regard to various considerations. The most positive gain is in the form of reduction of deflections, which is generally more significant for steel and wood structures because deflections are less often critical for concrete members. For the beam with one fixed end (Figure 4.3c), it may be noted that the value for maximum shear is increased and the maximum moment is the same as for the simple span (no gain in those regards). For full end fixity (Figure 4.3d), the shear is unchanged, while both moment and deflection are quite substantially reduced in magnitude.

For the rigid frames shown in Figure 4.2, the restraints will reduce moment and deflection for the beam, but the cost is at the expense of the columns, which must take some moment in addition to axial force. Rigid frames are often utilized to resist lateral loads due to wind and earthquakes, presenting complex combinations of lateral and gravity loading, which must be investigated.

Effects of Concentrated Loads

Framing systems for roofs and floors often consist of series of evenly spaced beams that are supported by other beams placed at right angles to them. The beams supporting beams are, thus, subjected to a series of spaced, concentrated loads—the end reactions of the supported beams. The effects of a single such load at the center of a beam span are shown in Figures 4.3e to h. Two additional situations of evenly spaced concentrated loading are shown in Figure 4.4. When more than three such loads occur, it is usually adequate to consider the sum of the concentrated loads as a uniformly distributed load and to use the values given for Figures 4.3b to d.

Multiple Beam Spans

Figure 4.5 shows various loading conditions for a beam that is continuous through two equal spans. When continuous spans occur, it is usually necessary to give some consideration to the possibilities of partial beam loading, as shown in Figures 4.5b and d. It may be noted for Figure 4.5b that although there is less total load on the beam, the values for maximum positive moment and for shear at the free end are higher than for the fully loaded beam in Figure 4.5a. This condition of partial loading must be considered for *live loads* (people, furniture, snow, etc.). For design, the partial loading effects due to the live load must be combined with those produced by *dead load* (permanent weight of the construction) for the full action of the beam.

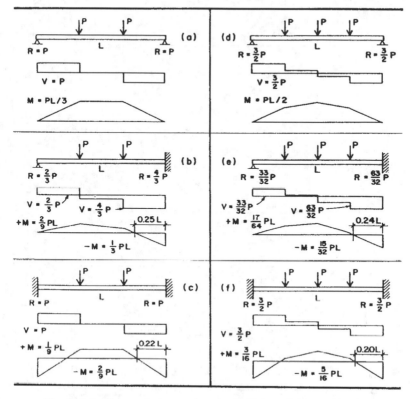

Figure 4.4 Beam response values for multiple concentrated loads.

Figure 4.6 shows a beam that is continuous through three equal spans, with various situations of uniform load on the beam spans. Figure 4.6a gives the loading condition for dead load (*always* present in *all* spans). Figures 4.6b to d show the several possibilities for partial loading, each of which produces some specific critical values for the reactions, shears, moments, and deflections.

Complex Loading and Span Conditions

Although values have been given for many common situations in Figures 4.3 through 4.6, there are numerous other possibilities in terms of unsymmetrical loadings, unequal spans, cantilevered free ends, and so on. Where these occur, an analysis of the indeterminate structure must be performed. For some additional conditions, the reader is referred to var-

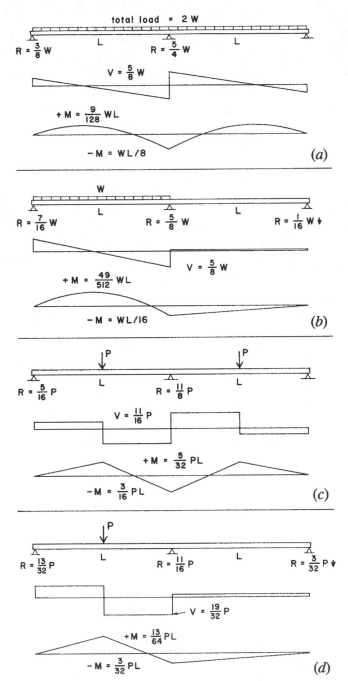

Figure 4.5 Response values for two-span beams.

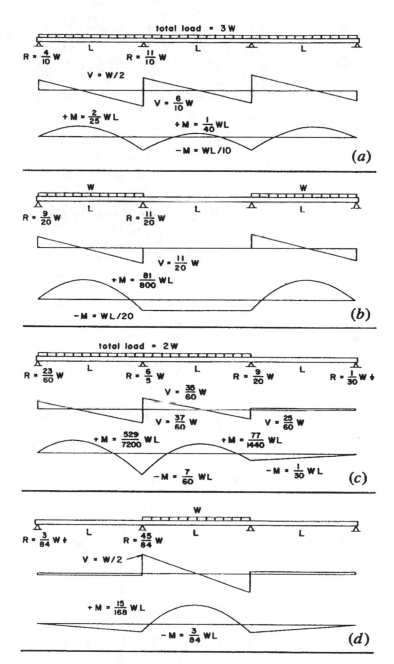

Figure 4.6 Response values for three-span beams.

ious handbooks that contain tabulations similar to those presented here. Two such references are the *CRSI Handbook* (Ref. 5) and the AISC Manual (*Manual of Steel Construction,* published by the American Institute of Steel Construction).

Loads and their required combinations for design are discussed in Chapter 15. Resisting loads is basically what structures exist to do. It is very important, therefore, to understand the source and derivation of loads for design, as well as the reliability of their quantification.

4.6   INVESTIGATION OF COLUMN AND BEAM FRAMES

Frames in which two or more of the members are attached to each other with connections that are capable of transmitting bending between the ends of the members are called *rigid frames.* The connections used to achieve such a frame are called *moment connections* or *moment-resisting connections.* Most rigid frame structures are statically indeterminate and do not yield to investigation by consideration of static equilibrium alone. The examples presented in this section are all rigid frames that have conditions that make them statically determinate and thus capable of being fully investigated by methods developed in this book.

Cantilever Frames

Consider the frame shown in Figure 4.7*a,* consisting of two members rigidly joined at their intersection. The vertical member is fixed at its base, providing the necessary support condition for stability of the frame. The horizontal member is loaded with a uniformly distributed loading and functions as a simple cantilever beam. The frame is described as a cantilever frame because of the single fixed support. The five sets of figures shown in Figures 4.7*b* through *f* are useful elements for the investigation of the behavior of the frame. They consist of the following:

1. The free body diagram of the entire frame, showing the loads and the components of the reactions (Figure 4.7*b*). Study of this figure will help in establishing the nature of the reactions and in determining the conditions necessary for stability of the frame as a whole.
2. The free body diagrams of the individual elements (Figure 4.7*c*). These are of great value in visualizing the interaction of the parts of the frame. They are also useful in the computations for the internal forces in the frame.

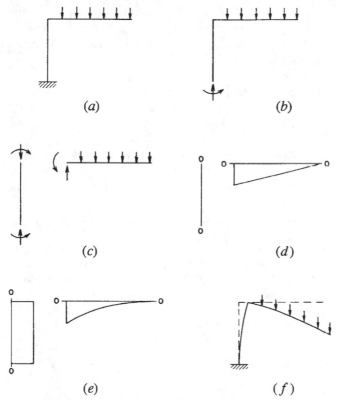

Figure 4.7 Diagrams for investigation of the rigid frame.

3. The shear diagrams of the individual elements (Figure 4.7d). These are sometimes useful for visualizing, or for actually computing, the variations of moment in the individual elements. No particular sign convention is necessary unless in conformity with the sign used for moment.

4. The moment diagrams for the individual elements (Figure 4.7e). These are very useful, especially in the determination of the deformation of the frame. The sign convention used is that of plotting the moment on the compression side of the element.

5. The deformed shape of the loaded frame (Figure 4.7f). This is the exaggerated profile of the bent frame, usually superimposed on an outline of the unloaded frame for reference. This is very useful for the general visualization of the frame behavior. It is particularly

useful for determination of the character of the external reactions and the form of interaction between the parts of the frame. Correlation between the deformed shape and the form of the moment diagram is a useful check.

When investigations are performed, these elements are not usually produced in the sequence just described. In fact, it is generally recommended that the deformed shape be sketched first so that its correlation with other factors in the investigation may be used as a check on the work. The following examples illustrate the process of investigation for simple cantilever frames:

Example 1.   Find the components of the reactions, and draw the free body diagrams, shear and moment diagrams, and the deformed shape of the frame shown in Figure 4.8a.

Solution: The first step is the determination of the reactions. Considering the free body diagram of the whole frame (Figure 4.8b), we compute the reactions as follows:

$$\Sigma F = 0 = +8 - R_v, \ R_v = 8 \text{ kips (up)}$$

and with respect to the support,

$$\Sigma M = 0 = M_R - (8 \times 4), \ M_R = 32 \text{ kip-ft (counterclockwise)}$$

Note that the sense, or sign, of the reaction components is visualized from the logical development of the free body diagram.

Consideration of the free body diagrams of the individual members will yield the actions required to be transmitted by the moment connection. These may be computed by application of the conditions for equilibrium for either of the members of the frame. Note that the sense of the force and moment is opposite for the two members, simply indicating that what one does to the other is the opposite of what is done to it.

In this example, there is no shear in the vertical member. As a result, there is no variation in the moment from the top to the bottom of the member. The free body diagram of the member, the shear and moment diagrams, and the deformed shape should all corroborate this fact. The shear and moment diagrams for the horizontal member are simply those for a cantilever beam.

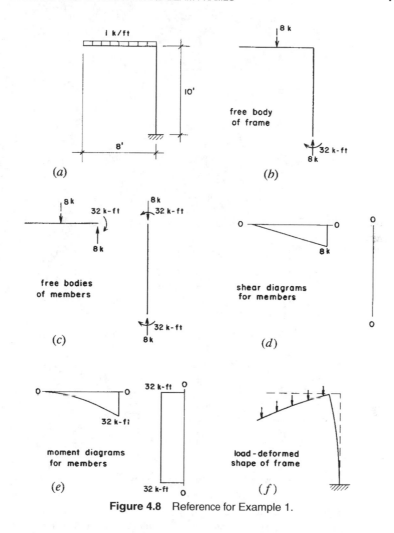

Figure 4.8 Reference for Example 1.

It is possible with this example, as with many simple frames, to visualize the nature of the deformed shape without recourse to any mathematical computations. It is advisable to do so as a first step in investigation and to check continually during the work that individual computations are logical with regard to the nature of the deformed structure.

Example 2. Find the components of the reactions and draw the shear and moment diagrams and the deformed shape of the frame in Figure 4.9a.

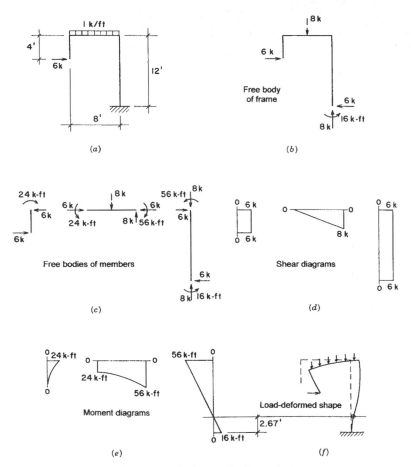

Figure 4.9 Reference for Example 2.

Solution: In this frame, there are three reaction components required for stability because the loads and reactions constitute a general coplanar force system. If the free body diagram of the whole frame (Figure 4.9*b*) is used, the three conditions for equilibrium for a coplanar system are used to find the horizontal and vertical reaction components and the moment component. If necessary, the reaction force components could be combined into a single-force vector, although this is seldom required for design purposes.

Note that the inflection occurs in the larger vertical member because the moment of the horizontal load about the support is greater than that of the vertical load. In this case, you must compute this value before you can draw the deformed shape accurately.

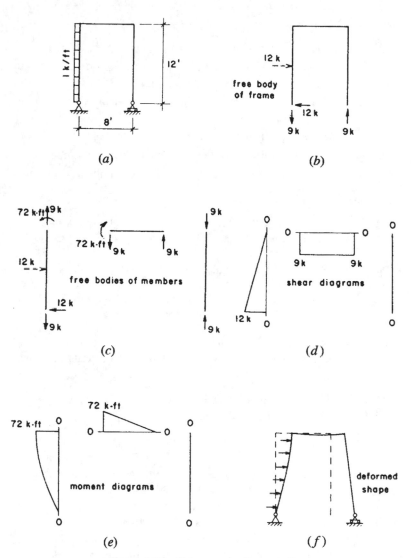

Figure 4.10 Reference for Example 3.

Example 3. Investigate the frame shown in Figure 4.10 for the reactions and internal conditions. Note that the right-hand support allows for an upward vertical reaction only, whereas the left-hand support allows for vertical and horizontal components. Neither support provides moment resistance.

Solution: Do the following:

1. Sketch the deflected shape (a little tricky but a good exercise).
2. Consider the equilibrium of the free body diagram of the whole frame to find the reactions.
3. Consider the equilibrium of the left-hand vertical member to find the internal actions at its top.
4. Consider the equilibrium of the horizontal member.
5. Consider the equilibrium of the right-hand vertical member.
6. Draw the shear and moment diagrams for the members. Then check to see that the required correlation exists.

Note: Before attempting the following exercise problems, you should produce the results shown in Figure 4.10 independently.

Problems 4.6.A, B, C.   For the frames shown in Figures 4.11a through c, find the components of the reactions, draw the free body diagrams of the whole frame and the individual members, draw the shear and moment diagrams for the individual members, and sketch the deformed shape of the loaded structure.

Problems 4.6.D, E.   Investigate the frames shown in Figures 4.11d and e for reactions and internal conditions, using the procedure shown for the preceding examples.

4.7 APPROXIMATE INVESTIGATION OF INDETERMINATE STRUCTURES

There are many possibilities for the development of rigid frames for building structures. Two common types of frames are the single-span bent and the vertical, planar bent, consisting of the multistory columns and multispan beams in a single plane in a multistory building.

As with other structures of a complex nature, the highly indeterminate rigid frame presents a good case for use of computer-aided methods. Programs utilizing the finite element method are available and are used frequently by professional designers. So-called *shortcut hand computation methods,* such as the moment distribution method were popular in the past. They are "shortcut" only in reference to more laborious hand computation methods; applied to a complex frame, they constitute a considerable effort—and then produce answers for only one loading condition.

Rigid-frame behavior is much simplified when the joints of the frame are not displaced; that is, they move only by rotating. This is usually true

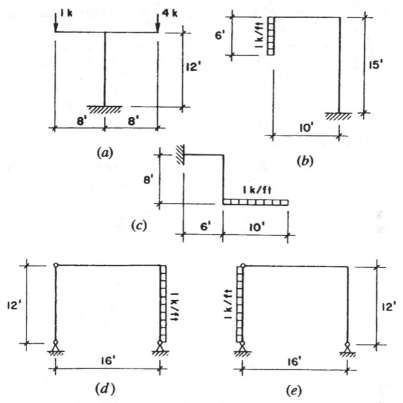

Figure 4.11   Reference for Problem 4.6.

for only the case of gravity loading on a symmetrical frame—and with only a symmetrical gravity load. If the frame is not symmetrical, the load is nonuniformly distributed, or lateral loads are applied, frame joints will move sideways (called *sidesway* of the frame) and additional forces will be generated by the joint displacements.

If joint displacement is considerable, there may be significant increases in force effects in vertical members due to the *P*-delta effect. (See Section 10.5.) In relatively stiff frames, with quite heavy members, this is usually not critical. In a highly flexible frame, however, the effects may be serious. In this case, the actual lateral movements of the joints must be computed to obtain the eccentricities used for determination of the *P*-delta effect. (See Section 10.12.) Reinforced concrete frames are typically quite stiff, so this effect is often less critical than for more flexible frames of wood or steel.

Lateral deflection of a rigid frame is related to the general stiffness of the frame. When several frames share a loading, as in the case of a multi-story building with several bents, the relative stiffnesses of the frames must be determined. This is done by considering their relative deflection resistances.

The Single-Span Bent

Figure 4.12 shows two possibilities for a rigid frame for a single-span bent. In Figure 4.12a, the frame has pinned bases for the columns, result-

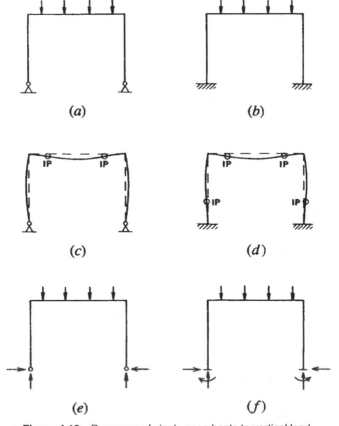

Figure 4.12   Response of single-span bents to vertical load.

ing in the load-deformed shape shown in Figure 4.12c, and the reaction components as shown in the free body diagram for the whole frame in Figure 4.12e. The frame in Figure 4.12b has fixed bases for the columns, resulting in the slightly modified behavior indicated. These are common situations: the base condition depending on the supporting structure as well as the frame itself.

The frames in Figure 4.12 are technically not statically determinate and require analysis by something more than statics. However, if the frame is symmetrical and the loading is uniform, the upper joints do not move sideways and the behavior is of a classic form. For this condition, analysis by moment area, three-moment equation, or moment distribution may be performed, although tabulated values for behaviors can also be obtained for this common form of structure.

Figure 4.13 shows the single-span bent under a lateral load applied at the upper joint. In this case, the upper joints move sideways, the frame taking the shape indicated with reaction components as shown. This also

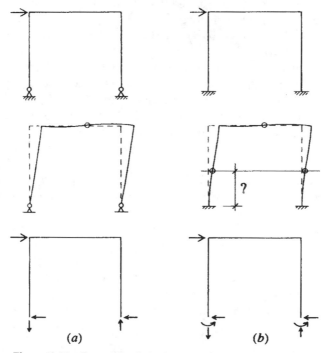

(a)   (b)

Figure 4.13 Response of single-span bents to horizontal load.

presents a statically indeterminate situation, although some aspects of the solution may be evident. For the pinned base frame in Figure 4.13, for example, a moment equation about one column base will cancel out the vertical reaction at that location, plus the two horizontal reactions, leaving a single equation for finding the value of the other vertical reaction. Then if the bases are considered to have equal resistance, the horizontal reactions will each simply be equal to one-half of the load. The behavior of the frame is, thus, completely determined, even though it is technically indeterminate.

For the frame with fixed column bases in Figure 4.13b, we may use a similar procedure to find the value of the direct force components of the reactions. However, the value of the moment at the fixed base is not subject to such simplified procedures. For this investigation of this frame—as well as that of the frames in Figure 4.12—it is necessary to consider the relative stiffness of the members, as is done in the moment distribution method or in any method for solution of the indeterminate structure.

The rigid-frame structure occurs quite frequently as a multilevel, multispan bent, constituting part of the structure for a multistory building. In most cases, such a bent is used as a lateral bracing element; although once it is formed as a moment-resistive framework, it will respond as such for all types of loads.

The multistory rigid bent is quite indeterminate, and its investigation is complex—requiring considerations of several different loading combinations. When loaded or formed unsymmetrically, it will experience sideways movements that further complicate the analysis for internal forces. Except for very early design approximations, the analysis is now sure to be done with a computer-aided system. The software for such a system is quite readily available.

For preliminary design purposes, it is sometimes possible to use approximate analysis methods to obtain member sizes of reasonable accuracy. Actually, many of the older high-rise buildings still standing were completely designed with these techniques—a reasonable testimonial to their effectiveness. A demonstration of approximate methods is given in Section 16.16.

Various design concerns for reinforced concrete rigid frames are discussed in Chapter 11.

5

LOAD AND RESISTANCE
FACTOR DESIGN (LRFD)

The general procedure for use of strength methods in design is currently described as the *load and resistance factor design method,* referred to for brevity as the *LRFD Method.* The basic concepts of strength design are described in Section 4.4. This chapter presents materials relating to the implementation of strength design by current methods. The procedures and data for this work come largely from the ACI Code.

5.1 LIMIT STATES VERSUS SERVICE CONDITIONS

The differences between the stress method and strength method were discussed in Chapter 4. Stress methods emphasize behavior at service load conditions, while strength methods relate primarily to the limits of resistance of structures. The service (actual anticipated usage) condition is not ignored in strength design. Some service behavior must be considered, such as that pertaining to deflections and development of cracks at service load levels. In addition, the service load is visualized as accurately as possible since it serves as the basis for derivation of the factored load.

A primary issue for strength design, however, is definition of limits for the behavior of the reinforced concrete structure. Some critical limits are the following:

1. *Yielding of the Reinforcing Steel.* This is a clear limit for the load capacity of the reinforcement. At this stress level, the extensive deformation of the steel will precipitate various forms of failure of the concrete: not necessarily collapse of the structure, but a shift in failure mechanisms.
2. *Excessive Cracking of the Concrete.* Development of stress in the steel will cause inevitable, unavoidable cracking; thus, it, becomes a matter of the degree of cracking. A major limit for design is the established limit for cracking for various situations.
3. *Compression Failure of the Concrete.* Most design procedures control the reinforced members so that initial failures occur by yielding of the tension steel. In beams, this ordinarily rules against compression failure as the first stage. However, in columns and doubly reinforced beams, the behavior is more complex. A special problem is that of providing for containment of the concrete, which is discussed in Chapters 10 and 11.

In fact, the LRFD method as currently implemented combines elements of the old stress method with newer elements of limit states analyses and risk analysis for a total design procedure. The objective is to kill two structural birds with one stone: make it work for service conditions, and make it safe by intelligent evaluation of its limiting capacity.

5.2 LOADS FOR DESIGN

Loads on structures derive from various sources, the primary ones being gravity, wind, and earthquakes. For use in investigation or design work, loads must first be identified, measured, or quantified in some way, and then—for the strength method—factored. In most situations, they must also be combined in all possible ways that are statistically likely, which typically produces more than one load condition for design. See the discussion in Section 15.8.

This is not the whole story for many structures because of special concerns. For example, the stability of a shear wall may be critical with a combination of only dead load and the lateral load—wind or seismic. Long-term stress conditions in wood or effects of creep in concrete may be critical with only dead load as a permanent load condition. In the end, good engineering design judgment must prevail to visualize the really necessary combinations.

A single loading combination may prevail for the consideration of maximum effect on a given structure. However, in complex structures (trusses, moment-resistive building frames, etc.), separate individual members may be designed for different critical load combinations. While the critical combination for simple structures can sometimes be easily visualized, it is sometimes necessary to perform complete investigations for many combinations and then to compare the results in detail to ascertain the true design requirements.

Loads used for the design of building structures, and their incorporation in the design process, are discussed in Chapter 15. The discussion in Chapter 15 also treats the use of load factors and required combinations of loads.

5.3 RESISTANCE FACTORS

Factoring (modifying) the loads is one form of adjustment for control of safety in strength design. The second basic adjustment is modifying the quantified resistance of the structure. This amounts to first determining its strength in some terms (compression resistance, moment capacity, buckling limit, etc.) and then reducing it by some percentage. The reduction percentage (obtained by using the resistance factor) is based on various considerations, including concerns for the reliability of theories, quality control in production, ability to accurately predict behaviors, and so on.

Strength design usually consists of comparing the factored load (the load increased by some percentage) to the factored resistance (the resistance reduced by some percentage) of the loaded structure. The factored resistance, called the *design strength,* is obtained by using the following values for the resistance factor, ϕ:

- 0.90 for flexure, axial tension, and combinations of flexure and tension
- 0.75 for columns with spirals
- 0.70 for columns with ties
- 0.85 for shear and torsion
- 0.70 for compression bearing

- 0.65 for other reinforced members
- 0.55 for plain (not reinforced) concrete

When combined with the load factors, application of the resistance factors amounts to a magnification of the safety percentage level.

5.4 STRENGTH DESIGN PROCESSES

Application of design procedures in the working stress method tend to be simpler and more direct than in the strength methods. For example, the design of a beam may amount to simply inverting a few basic stress or strain equations to derive some required properties (section modulus for bending, area for shear, moment of inertia for deflection, etc.). Strength method applications tend to be more obscure, mostly because the mathematical formulations and data are usually more complex and less direct.

Extensive experience with either method will eventually produce some degree of intuitive judgment, allowing designers to make quick approximations even before the derivation of any specific requirements. Thus, an experienced designer can look merely at the basic form, dimensions, and general usage of a structure and quickly determine an approximate solution—probably quite close in most regards to what a highly exact investigation will produce. Having designed many similar structures before is the essential basis for this quick solution. This can be useful when some error in computation causes the exact design process to produce a weird answer.

Extensive use of strength methods has required careful visualizations and analyses of modes of failures. After many such studies, one develops some ability to quickly ascertain the single, major response characteristic that is the primary design determinant for a particular structure. Concentration on those selected, critical design factors helps to quickly establish a reasonable design, which can then be tested by many basically routine investigative procedures for other responses.

Use of multiple load combinations, multimode structural failures, multiple stress and strain analyses, and generally complex investigative or design procedures is much assisted by computers in most professional design work. A treacherous condition, however, is to be a slave to the computer, accepting its answers with no ability to judge their true appropriateness or correctness. Grinding it out by hand—at least a few times—helps one to appreciate the process.

6

REINFORCED CONCRETE
FLEXURAL MEMBERS

In concrete structures subjected to bending, two kinds of internal resistance are required. The concrete can often develop the compression component required but needs help in developing tension resistance. As a result, it is common practice to add steel reinforcement to flexural members for tension resistance. This chapter treats the various concerns for the design of reinforced concrete members subjected to bending.

6.1 GENERAL FLEXURAL ACTIONS

The primary concerns for beams relate to their necessary resistance to bending and shear and some limitations on their deflection. For wood or steel beams, the usual concern is only for the singular maximum values of bending and shear in a given beam. For concrete beams, on the other hand, it is necessary to provide for the values of bending and shear as they vary along the entire length of a beam—even through multiple spans in the case of continuous beams, which are a common occurrence in concrete struc-

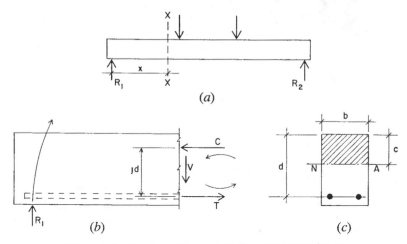

Figure 6.1 Bending action in a reinforced concrete beam.

tures. For simplification of the work, it is necessary to consider the actions of a beam at a specific location in the beam, but one must bear in mind that this action must be integrated with all the other effects on the beam throughout its length.

When a member is subjected to bending, such as the beam shown in Figure 6.1a, internal resistances of two basic kinds are generally required. Internal actions are "seen" by visualizing a cut section, such as that taken at X-X in Figure 6.1a. When the portion of the beam to the left of the cut section is removed, its free body actions are as shown in Figure 6.1b. At the cut section, consideration of static equilibrium requires the development of the internal shear force (V in the figure) and the internal resisting moment (represented by the force couple: C and T in the figure).

6.2 BEHAVIOR OF REINFORCED CONCRETE BEAMS

If a beam consists of a simple rectangular concrete section with tension reinforcement only, as shown in Figure 6.1c, the force C is considered to be developed by compressive stresses in the concrete—indicated by the shaded area above the neutral axis. The tension force, however, is considered to be developed by the steel alone, ignoring the tensile resistance of the concrete. For low-stress conditions the latter is not true, but at a greater level of stress the tension-weak concrete will indeed crack, virtually leaving the steel unassisted, as assumed.

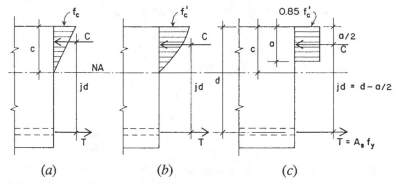

Figure 6.2 Development of bending stress actions in a reinforced concrete beam.

At moderate levels of stress, the resisting moment is visualized as shown in Figure 6.2*a*, with a linear variation of compressive stress from zero at the neutral axis to a maximum value of f_c at the edge of the section. As stress levels increase, however, the nonlinear stress-strain character of the concrete becomes more significant, and it becomes necessary to acknowledge a more realistic form for the compressive stress variation, such as that shown in Figure 6.2*b*. As stress levels approach the limit of the concrete, the compression becomes vested in an almost constant magnitude of unit stress, concentrated near the top of the section. For strength design, in which the moment capacity is expressed at the ultimate limit, it is common to assume the form of stress distribution shown in Figure 6.2*c*, with the limit for the concrete stress set at 0.85 times f'_c. Expressions for the moment capacity derived from this assumed distribution compare reasonably with the response of beams tested to failure in laboratory experiments.

Response of the steel reinforcement is more simply visualized and expressed. Since the steel area in tension is concentrated at a small location with respect to the size of the beam, the stress in the bars is considered to be a constant. Thus, at any level of stress, the total value of the internal tension force may be expressed as

$$T = A_s f_s$$

and for the practical limit of *T*,

$$T = A_s f_y$$

6.3 INVESTIGATION AND DESIGN FOR FLEXURE

The following is a presentation of the formulas and procedures used in the strength method. The discussion is limited to a rectangular beam section with tension reinforcement only.

Referring to Figure 6.3, the following are defined:

- b = width of the concrete compression zone
- d = effective depth of the section for stress analysis; from the centroid of the steel to the edge of the compressive zone
- h = overall depth (height) of the section
- A_s = cross-sectional area of reinforcing bars
- p = percentage of reinforcement, defined as

$$p = \frac{A_s}{bd}$$

Figure 6.2c shows the rectangular "stress block" that is used for analysis of the rectangular section with tension reinforcing only by the strength method. This is the basis for investigation and design as provided for in the ACI Code.

The rectangular stress block is based on the assumption that a concrete stress of $0.85\,f'_c$ is uniformly distributed over the compression zone, which has dimensions equal to the beam width b and the distance

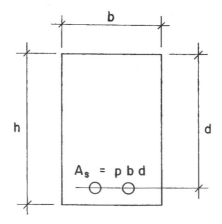

Figure 6.3   Reference notation for a reinforced concrete beam section.

a, which locates a line parallel to and above the neutral axis. The value of a is determined from the expression $a = \beta_1 \times c$, where β_1 (beta one) is a factor that varies with the compressive strength of the concrete, and c is the distance from the extreme fiber to the neutral axis (see Figure 6.2c). For concrete having f'_c equal to or less than 4000 psi [27.6 MPa], the ACI Code gives a maximum value for $a = 0.85\ c$.

With the rectangular stress block, the magnitude of the compressive force in the concrete is expressed as

$$C = (0.85\, f'_c)(b)(a)$$

and it acts at a distance of $a/2$ from the top of the beam. The arm of the resisting force couple then becomes $d - (a/2)$, and the developed resisting moment as governed by the concrete is

$$M_c - C\left(d - \frac{a}{2}\right) = 0.85\, f'_c b a\left(d - \frac{a}{2}\right) \tag{6.3.1}$$

With T expressed as $A_s \times f_y$, the developed moment as governed by the reinforcement is

$$M_t = T\left(d - \frac{a}{2}\right) = A_s f_y\left(d - \frac{a}{2}\right) \tag{6.3.2}$$

A formula for the dimension a of the stress block can be derived by equating the compression and tension forces; thus,

$$0.85\, f'_c b a = A_s f_y, \qquad a = \frac{A_s f_y}{0.85\, f'_c b} \tag{6.3.3}$$

By expressing the area of steel in terms of a percentage p, the formula for a may be modified as follows:

$$p = \frac{A_s}{bd}, \qquad A_s = pbd$$

$$a = \frac{(pbd)f_y}{0.85\, f'_c b} = \frac{pdf_y}{0.85\, f'_c} \quad \text{or} \quad \frac{a}{d} = \frac{pf_y}{0.85\, f'_c} \tag{6.3.4}$$

A useful reference is the so-called *balanced section,* which occurs when using the exact amount of reinforcement results in the simultaneous development of the limiting stresses in the concrete and steel. The balanced section for strength design is visualized in terms of strain rather than stress. The limit for a balanced section is expressed in the form of the percentage of steel required to produce balanced conditions. The formula for this percentage is

$$p_b = \frac{0.85f'_c}{f_y} \times \frac{87}{87 + f_y} \qquad (6.3.5)$$

in which f'_c and f_y are in units of ksi.

Returning to the formula for the developed resisting moment, as expressed in terms of the steel, we can derive a useful formula as follows:

$$M_t = A_s f_y \left(d - \frac{a}{2} \right)$$

$$= (pbd)(f_y)\left(d - \frac{a}{2} \right)$$

$$= (pbd)(f_y)(d)\left(1 - \frac{a}{2d} \right)$$

$$= (bd^2)\left[pf_y\left(1 - \frac{a}{2d} \right) \right]$$

Thus,

$$M_t = Rbd^2 \qquad (6.3.6)$$

where

$$R = pf_y\left(1 - \frac{a}{2d} \right) \qquad (6.3.7)$$

With the reduction factor applied, the design moment for a section is limited to nine-tenths of the theoretical resisting moment.

Values for the balanced section factors (p, R, and a/d) are given in Table 6.1 for various combinations of f'_c and f_y. The balanced section is not necessarily a practical one for design. In most cases, economy will be

TABLE 6.1   Balanced Section Properties for Rectangular Sections with Tension Reinforcement Only

f_y		f'_c				R	
ksi	MPa	ksi	MPa	p	a/d	ksi	kPa
40	276	2	13.8	0.0291	0.685	0.766	5280
		3	20.7	0.0437	0.685	1.149	7920
		4	27.6	0.0582	0.685	1.531	10600
		5	34.5	0.0728	0.685	1.914	13200
60	414	2	13.8	0.0168	0.592	0.708	4890
		3	20.7	0.0252	0.592	1.063	7330
		4	27.6	0.0335	0.592	1.417	9770
		5	34.5	0.0419	0.592	1.771	12200

achieved by using less than the balanced reinforcing for a given concrete section. In special circumstances it may also be possible, or even desirable, to use compressive reinforcing in addition to tension reinforcing. Nevertheless, the balanced section is often a useful reference when design is performed.

Beams with reinforcement less than that required for the balanced moment are called *underbalanced sections* or *under-reinforced sections*. If a beam must carry bending moment in excess of the balanced moment for the section, it is necessary to provide some compressive reinforcement, as discussed in Section 6.6. The balanced section is not necessarily a design ideal, but it is useful in establishing the limits for the section.

In the design of concrete beams, two situations commonly occur. The first occurs when the beam is entirely undetermined; that is, the concrete dimensions and the reinforcement are unknown. The second occurs when the concrete dimensions are given, and the required reinforcement for a specific bending moment must be determined. The following examples illustrate the use of the formulas just developed for each of these problems:

Example 1.   The service load bending moments on a beam are 58 kip-ft [78.6 kN-m] for dead load and 38 kip-ft [51.5 kN-m] for live load. The beam is 10 in. [254 mm] wide, f'_c is 3000 psi [20.7 MPa], and f_y is 60 ksi [414 MPa]. Determine the depth of the beam and the tensile reinforcing required.

Solution: The first step is to determine the required moment, using the load factors. Thus,

$$U = 1.2(D) + 1.6(L)$$

$$M_u = 1.2(M_{DL}) + 1.6(M_{LL})$$

$$= 1.2(58 \text{ kip-ft}) + 1.6(38 \text{ kip-ft})$$

$$= 130.4 \text{ kip-ft } [177 \text{ kN-m}]$$

With the capacity reduction of 0.90 applied, the desired moment capacity of the section is determined as

$$M_t = \frac{M_u}{0.90} = \frac{130.4 \text{ kip-ft}}{0.90} = 145 \text{ kip-ft}$$

$$= 145 \text{ kip-ft} \times \left(\frac{12 \text{ in.}}{1 \text{ ft}} \right) = 1740 \text{ kip-in. } [197 \text{ kN-m}]$$

The reinforcement ratio as given in Table 6.1 is $p = 0.0252$. The required area of reinforcement for this section may, thus be, determined from the relationship

$$A_s = pbd$$

While there is nothing especially desirable about a balanced section, it does represent the beam section with least depth if tension reinforcing only is used. Therefore, proceed to find the required balanced section for this example.

To determine the required effective depth *d,* use Equation (6.3.6); thus,

$$M_t = Rbd^2$$

With the value of $R = 1.063$ ksi from Table 6.1,

$$M_1 = 1740 \text{ kip-in.} = 1.063 \text{ ksi } (10 \text{ in.})(d)^2$$

and

$$d = \sqrt{\frac{M}{Rb}} = \sqrt{\frac{1740 \text{ kip-in.}}{1.063 \text{ ksi } (10 \text{ in.})}} = \sqrt{164 \text{ in.}^2} = 12.8 \text{ in. } [325 \text{ mm}]$$

If this value is used for d, the required steel area may be found using the value of $p = 0.0252$ from Table 6.1; thus,

$$A_s = pbd = 0.0252(10 \text{ in.})(12.8 \text{ in.}) = 3.23 \text{ in.}^2 \text{ [2084 mm}^2]$$

Although they are not given in this example, there are often some considerations other than flexural behavior alone that influence the choice of specific dimensions for a beam. These may include:

- Design for shear
- Coordination of the depths of a set of beams in a framing system
- Coordination of the beam dimensions and placement of reinforce ment in adjacent beam spans
- Coordination of beam dimensions with supporting columns
- Limiting beam depth to provide overhead clearance beneath the structure

If the beam is of the ordinary form shown in Figure. 6.4, the specified dimension is usually that given as h. Assuming the use of a No. 3 U-stirrup, a cover of 1.5 in. [38 mm], and an average-size reinforcing bar of 1-in. [25-mm] diameter (No. 8 bar), the design dimension d will be less than h by 2.375 in. [60 mm]. Lacking other considerations, the overall depth of the beam (h) will be 16 in.

Next, select a set of reinforcing bars to obtain this area. For the pur pose of the example, select bars all of a single size (see Table 1.1); the number required will be:

- No. 6 bars: 3.23/0.44 = 7.3, or 8 [2084/284 = 7.3]
- No. 7 bars: 3.23/0.60 − 5.4, or 6 [2084/387 − 5.4]
- No. 8 bars: 3.23/0.79 = 4.1, or 5 [2084/510 = 4.1]
- No. 9 bars: 3.23/1.00 = 3.3, or 4 [2084/645 = 3.3]
- No. 10 bars: 3.23/1.27 = 2.5, or 3 [2084/819 = 2.5]
- No. 11 bars: 3.23/1.56 = 2.1, or 3 [2084/1006 = 2.1]

In real design situations, there are always various additional considerations that influence the choice of the reinforcing bars. One general desire is that of having the bars in a single layer, as this keeps the centroid of the steel as close as possible to the edge (bottom in this case) of the member, giving the greatest value for d with a given height (h) of a concrete section. With the section as shown in Figure 6.4, a beam width of 10 in. will

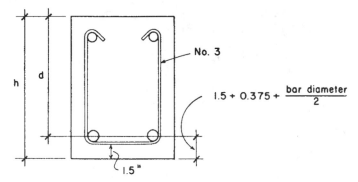

Figure 6.4 Common form of a reinforced concrete beam.

yield a net width of 6.25 in. inside the No. 3 stirrups (outside width of 10 less 2 × 1.5 cover and 2 × 0.375 stirrup diameter). Applying the code criteria for minimum spacing for this situation, the required width for the various bar combinations can be determined. Minimum space required between bars is one bar diameter or a lower limit of one inch. (See discussion in Section 2.6.) Two examples for this are shown in Figure 6.5. It will be found that none of the choices will fit this beam width. Thus the beam width must be increased or two layers of bars must be used.

If there are reasons, as there often are, for not selecting the least deep section with the greatest amount of reinforcing, a slightly different procedure must be used, as illustrated in the following example:

Example 2. Using the same data as in Example 1, find the reinforcement required if the desired beam section has $b = 10$ in. [254 mm] and $d = 18$ in. [457 mm].

Solution: The first two steps in this situation would be the same as in Example 1—to determine M_u and M_r. The next step would be to determine whether the given section is larger than, smaller than, or equal to a balanced section. Since this investigation has already been done in Example 1, observe that the 10 in. by 18 in. section is larger than a balanced section. Thus, the actual value of a/d will be less than the balanced section value of 0.592. The next step would then be as follows:

Estimate a value for a/d—something smaller than the balanced value. For example, try $a/d = 0.3$. Then

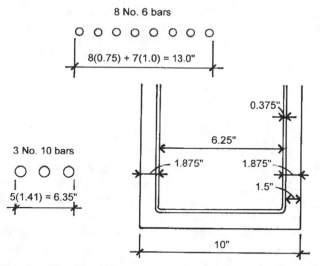

Figure 6.5 Consideration of beam width for proper spacing of a single layer of reinforcing bars.

$$a = 0.3d = 0.3(18 \text{ in.}) = 5.4 \text{ in. } [137 \text{ mm}]$$

With this value for *a*, use Equation 6.2.2 to find a required value for A_s. Referring to Figure 6.2,

$$M_t = T(jd) = (A_s f_y)\left(d - \frac{a}{2}\right)$$

$$A_s = \frac{M_t}{f_y\left(d - \dfrac{a}{2}\right)} = \frac{1740 \text{ kip-in.}}{60 \text{ ksi } (15.3 \text{ in.})} = 1.89 \text{ in.}^2 \ [1220 \text{ mm}^2]$$

Next, test to see if the estimate for *a/d* was close by finding *a/d* using Equation 6.2.4. Thus,

$$p = \frac{A_s}{bd} = \frac{1.89 \text{ in.}^2}{10 \text{ in. } (18 \text{ in.})} = 0.0105$$

and, from Equation 6.2.4

$$\frac{a}{d} = \frac{pf_y}{0.85f'_c} = \frac{0.0105\ (60\ \text{ksi})}{0.85\ (3\ \text{ksi})} = 0.247$$

Thus,

$$a = 0.247\ (18\ \text{in.}) = 4.45\ \text{in.}, \qquad d - \frac{a}{2} = 15.8\ \text{in.}\ [399\ \text{mm}]$$

If this value for $d - a/2$ is used to replace that used earlier, the required value of A_s will be slightly reduced. In this example, the correction will be only a few percent. If the first guess of a/d was way off, it might justify another run through the analysis to get closer to an exact answer.

For beams that are classified as under-reinforced (section dimensions larger than the limit for a balanced section), check for the minimum required reinforcement. For a rectangular section, the ACI Code specifies that a minimum area be

$$A_s = \frac{3\sqrt{f'_c}}{f_y}(bd)$$

but not less than

$$A_s = \frac{200}{f_y}(bd)$$

On the basis of these requirements, values for minimum reinforcement for rectangular sections with tension reinforcement only are given in Table 6.2 for two grades of steel and three concrete strengths.

TABLE 6.2   Minimum Required Tension Reinforcement for Rectangular Sections[a]

f'_c (psi)	$f_y = 40$ ksi	$f_y = 60$ ksi
3000	0.0050	0.00333
4000	0.0050	0.00333
5000	0.0053	0.00354

[a]Required A_s equals table value times bd of the beam section.

For the example, with a concrete strength of 3000 psi and f_y of 60 ksi, the minimum area of steel is, thus,

$$A_s = 0.00333(bd) = 0.00333(10 \times 18) = 0.60 \text{ in.}^2 \text{ [387 mm}^2]$$

which is clearly not critical in this case.

Problem 6.3.A. A rectangular concrete beam has $f'_c = 3000$ psi [20.7 MPa] and steel with $f_y = 40$ ksi [276 MPa]. Select the beam dimensions and reinforcement for a balanced section if the beam sustains a moment due to dead load of 60 kip-ft [81.4 kN-m] and a moment due to live load of 90 kip-ft [122 kN-m].

Problem 6.3.B. Same as Problem 6.3.A, except $f'_c = 4000$ psi [27.6 MPa], $f_y = 60$ ksi [414 MPa], $M_{DL} = 36$ kip-ft [48.8 kN-m], and $M_{LL} = 65$ kip-ft [88.1 kN-m].

Problem 6.3.C. Find the area of steel reinforcement required and select the bars for the beam in Problem 6.3.A if the section dimensions arc $b = 16$ in [406 mm] and $d = 32$ in [813 mm].

Problem 6.3.D. Find the area of steel reinforcement required and select the bars for the beam in Problem 6.3.B if the section dimensions are $b = 14$ in. [356 mm] and $d = 25$ in. [635 mm].

Use of Beam Tables

Compiling tables for design of concrete beams is complicated by the large number of combinations of concrete strength and steel yield strength. Limiting these values to those most commonly used may reduce the amount of tabulation. However, the number of possible combinations of beam dimensions is also extensive. Handbooks with beam design tables do exist, a widely used one being the *CRSI Handbook* (Reference 5). Professional structural designers use these types of references, as well as computer-aided processes.

Table 6.3 contains a limited number of beam examples with a range of values for beam width and effective depth. The table uses a single combination of strength values: 4 ksi for concrete strength and 60 ksi for steel yield stress. For each size of beam listed, four different choices for reinforcement are shown. The four choices for reinforcement are based on an assumption of a value for the ratio a/d. Specific combinations of bars mostly do not conform exactly to the a/d values but approximate them with practical bar choices. The percentage of steel area corresponding to

TABLE 6.3 Factored Moment Resistance of Concrete Beams, ϕM_r[1]

b × d (inches)	Approximate Values For a/d			
	0.1	0.2	0.3	0.4
	Approximate Values For p			
	0.00567	0.01133	0.0170	0.0227
10 × 14	2 #6 53	2 #8 90	3 #8 126	3 #9 151
10 × 18	3 #5 68	2 #9 146	3 #9 207	(3 #10) 247
10 × 22	2 #7 113	3 #8 211	(3 #10) 321	(3 #11) 371
12 × 16	2 #7 82	3 #8 154	4 #8 193	3 #11 270
12 × 20	2 #8 135	3 #9 243	4 #9 306	(2 #10+2-#11) 407
12 × 24	2 #8 162	3 #9 292	(4 #10) 466	(4 #11) 539
15 × 20	3 #7 154	4 #8 256	5 #9 382	(4 #11) 449
15 × 25	3 #8 253	4 #9 405	4 #11 597	(3 #10+3 #11) 764
15 × 30	3 #8 304	5 #9 607	(5 #11) 895	(3 #10+4 #11) 1085
18 × 24	3 #8 243	5 #9 486	6 #10 700	(6 #11) 809
18 × 30	3 #9 384	6 #9 729	(6 #11) 1074	(8 #11) 1348
18 × 36	3 #10 566	6 #10 1110	(7 #11) 1504	(9 #11) 1819
20 × 30	3 #10 489	7 #9 850	6 #11 1074	(9 #11) 1516
20 × 35	4 #9 598	5 #11 1106	(7 #11) 1462	(10 #11) 1966
20 × 40	6 #8 810	6 #11 1516	(9 #11) 2148	(12 #11) 2696
24 × 32	6 #8 648	7 #10 1152	(8 #11) 1527	(11 #11) 1977
24 × 40	6 #9 1026	7 #11 1769	(10 #11) 2387	(14 #11) 3145
24 × 48	5 #10 1303	(8 #11) 2426	(13 #11) 3723	(17 #11) 4583

[1] Table yields values of factored moment resistance in kip-ft with reinforcement indicated. Reinforcement choices shown in parentheses require greater width of beam or use of two stacked layers of bars. $f_c' = 4$ ksi, $f_y = 60$ ksi.

the *a/d* values are bracketed between the requirement for minimum reinforcement in Table 6.2 and the upper limit represented by the balanced section values in Table 6.1.

A practical consideration for the amount of reinforcement that can be used in a beam is that of the spacing required as related to the beam width. In Table 6.3, the bar combinations that cannot be accommodated in a single layer are indicated in parentheses.

For each combination of concrete dimensions and bar choices, Table 6.3 yields a value for the factored moment resistance of the beam. The following example illustrates a possible use for this table:

Example 3. Using Table 6.3, find acceptable choices of beam dimensions and reinforcement for a factored moment of 1000 kip-ft.

Solution: Possible choices from the table are:

- 15 × 30, 3 No. 10 + 4 No. 11 (requires two layers)
- 18 × 30, 6 No. 11 (requires two layers)
- 18 × 36, 6 No. 10
- 20 × 30, 6 No. 11
- 20 × 35, 5 No. 11
- 24 × 32, 7 No. 10
- 24 × 40, 6 No. 9

Using this range of possibilities, together with other design considerations for the beam size, a quick approximation for the beam design can be determined. A review of the work for Example 1 should indicate the practical use of this easily determined information.

Problems 6.3.E, F. Using Table 6.3, find acceptable choices for beam dimensions and reinforcement for a factored moment of: E, 400 kip-ft [542 kN-m]; F, 1200 kip-ft [1630 kN-m].

6.4 BEAMS IN SITECAST SYSTEMS

In sitecast construction, it is common to cast as much of the total structure as possible in a single, continuous pour. The length of the workday, the size of the available work crew, and other factors may affect this decision. Other considerations involve the nature, size, and form of the structure. For example, a convenient single-cast unit may consist of the whole floor structure for a multistory building, if it can be cast in a single workday.

Planning of the concrete construction is itself a major design task. The issue in consideration here is that such work typically results in the achieving of continuous beams and slabs, versus the common condition of simple-span elements in wood and steel construction. The design of continuous-span elements involves more complex investigation for behaviors due to the statically indeterminate nature of internal force resolution involving bending moments, shears, and deflections. For concrete structures, additional complexity results from the need to consider conditions all along the beam length, not just at locations of maximum responses.

In the upper part of Figure 6.6 the condition of a simple-span beam subjected to a uniformly distributed loading is shown. The typical bending moment diagram showing the variation of bending moment along the beam length takes the form of a parabola, as shown in Figure 6.6b. As with a beam of any material, the maximum effort in bending resistance must respond to the maximum value of the bending moment—here occurring at the midspan. For a concrete beam, the concrete section and its reinforcement must be designed for this moment value. However, for a long span and a large beam with a lot of reinforcement, it may be possible to reduce the amount of steel at points nearer to the beam ends. That is, some steel bars may be full length in the beam, while some others are only partial length and occur only in the midspan portion (see Figure 6.6c).

Figure 6.6d shows the typical situation for a continuous beam in a site-cast slab and beam framing system. For a single uniformly distributed loading, the moment diagram takes a form as shown in Figure 6.6e, with positive moments near the beam's midspan and negative moments at the supports. Locations for reinforcement that respond to the sign of these moments are shown on the beam elevation in Figure 6.6f.

For the continuous beam, it is obvious that separate requirements for the beam's moment resistance must be considered at each of the locations of peak values on the moment diagram. However, there are many additional concerns as well. Principal considerations include the following:

T-Beam Action.   At points of positive bending moment (midspan) the slab and beam monolithic construction must be considered to function together, giving a T-shaped form for the portion of the beam section that resists compression.

Use of Compression Reinforcement.   If the beam section is designed to resist the maximum bending moment, which occurs at only one point along the beam length, the section will be over-strong for all other locations. For this or other reasons, it may be advisable to use compres-

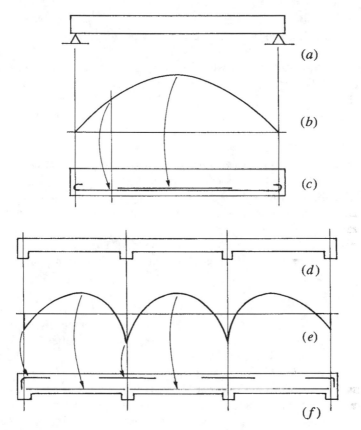

Figure 6.6 Utilization of reinforcement in concrete beams: (*a*) simple beam, (*b*) form of moment diagram for uniformly distributed loading on a simple beam, (*c*) use of reinforcement for a simple beam, (*d*) continuous beam, typical in concrete construction, (*e*) form of moment diagram for uniformly distributed loading on a continuous beam, and (*f*) use of reinforcement for a continuous beam.

sive reinforcement to reduce the beam size at the singular locations of maximum bending moment. This commonly occurs at support points and refers to the negative bending moment requiring steel bars in the top of the beam. At these points an easy way to develop compressive reinforcement is to simply extend the bottom steel bars (used primarily for positive moments) through the supports.

Spanning Slabs. Design of sitecast beams must usually be done in conjunction with the design of the slabs that they support. Basic considerations for the slabs are discussed in Section 6.7. The whole case for the slab-beam sitecast system is discussed in Chapter 9.

Beam Shear. While consideration for bending is a major issue, beams must also be designed for shear effects. This is discussed in Chapter 7. While special reinforcement is typically added for shear resistance, its interaction and coexistence in the beam with the flexural reinforcement must be considered.

Development of Reinforcement. This refers generally to the proper anchorage of the steel bars in the concrete so that their resistance to tension can be developed. At issue primarily are the exact locations of the end cutoffs of the bars and the details such as that for the hooked bar at the discontinuous end at the far left support shown in Figure 6.6f. It also accounts for the hooking of the ends of some of the bars in the simple beam. The general problems of bar development are discussed in Chapter 8.

6.5 T-BEAMS

When a floor slab and its supporting beams are cast at the same time, the result is monolithic construction in which a portion of the slab on each side of the beam serves as the flange of a T-beam. The part of the section that projects below the slab is called the *web* or *stem* of the T-beam. This type of beam is shown in Figure 6.7a. For positive moment, the flange is in compression and there is ample concrete to resist compressive stresses, as shown in Figure 6.7b or c. However, in a continuous beam, there are negative bending moments over the supports, and the flange here is in the tension stress zone with compression in the web. In this situation, the beam is assumed to behave essentially as a rectangular section with dimensions b_w and d, as shown in Figure 6.7d. This section is also used in determining resistance to shear. The required dimensions of the beam are often determined by the behavior of this rectangular section. What remains for the beam is the determination of the reinforcement required at the midspan where the T-beam action is assumed.

The effective flange width (b_f) to be used in the design of symmetrical T-beams is limited to one-fourth the span length of the beam. In addition, the overhanging width of the flange on either side of the web is limited to eight times the thickness of the slab or one-half the clear distance to the next beam.

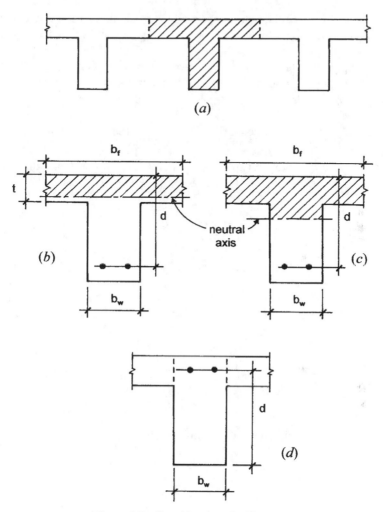

Figure 6.7 Considerations for T-beams.

In monolithic construction with beams and one-way solid slabs, the effective flange area of the T-beams is usually quite capable of resisting the compressive stresses caused by positive bending moments. With a large flange area, as shown in Figure 6.7a, the neutral axis of the section usually occurs quite high in the beam web. If the compression developed in the web is ignored, the net compression force may be considered to be located at the centroid of the trapezoidal stress zone that represents the stress distribution in the flange, and the compression force is located at something less than $t/2$ from the top of the beam.

An approximate analysis of the T-section by the strength method that avoids the need to find the location of the neutral axis and the centroid of the trapezoidal stress zone, consists of the following steps:

1. Determine the effective flange width for the T, as previously described.
2. Ignore compression in the web and assume a constant value for compressive stress in the flange (see Figure 6.8). Thus,

$$jd = d - \frac{t}{2}$$

Then, find the required steel area as

$$M_r = \frac{M_u}{0.9} = T(jd) = A_s f_y \left(d - \frac{t}{2} \right)$$

$$A_s = \frac{M_t}{f_y \left(d - \dfrac{t}{2} \right)}$$

3. Check the compressive stress in the concrete as

$$f_c = \frac{C}{b_f t} \leq 0.85 f'_c$$

where

$$C = \frac{M_r}{jd} = \frac{M_r}{d - \dfrac{t}{2}}$$

The value of maximum compressive stress will not be critical if this computed value is significantly less than the limit of $0.85 f'_c$.

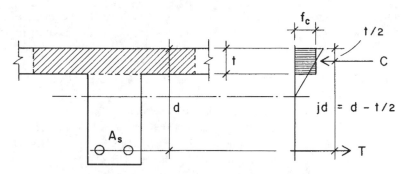

Figure 6.8 Basis for simplified analysis of a T-beam.

4. T-beams ordinarily function for positive moments in continuous beams. Because these moments are typically less than those at the beam supports, and the required section is typically derived for the more critical bending at the supports, the T-beam is typically considerably under reinforced. This makes it necessary to consider the problem of minimum reinforcement, as discussed for the rectangular section. The ACI Code provides special requirements for this for the T-beam, for which the minimum area required is defined as the greater value of

$$A_s = \frac{6\sqrt{f'_c}}{F_y}(b_w d)$$

or

$$A_s = \frac{3\sqrt{f'_c}}{F_y}(b_f d)$$

in which

- b_w = the width of the beam web
- b_f = the effective width of the T flange

The following example illustrates the use of this procedure. It assumes a typical design situation in which the dimensions of the section (b_f, b_w, d, and *t*—see Figure 6.7) are all predetermined by other design considerations and the design of the T-section is reduced to the requirement to determine the area of tension reinforcement.

Example 3. A T-section is to be used for a beam to resist positive moment. The following data is given: beam span = 18 ft [5.49 m], beams are 9 ft [2.74 m] center to center, slab thickness is 4 in. [102 mm], beam stem dimensions are b_w = 15 in. [381 mm] and d = 22 in. [559 mm], f'_c = 4 ksi [27.6 MPa], f_y = 60 ksi [414 MPa]. Find the required area of steel and select the reinforcing bars for a dead load moment of 125 kip-ft [170 kN-m] plus a live load moment of 100 kip-ft [136 kN-m].

Solution: Determine the effective flange width (necessary only for a check on the concrete stress). The maximum value for the flange width is

$$b_f = \frac{\text{span}}{4} = \frac{18 \times 12}{4} = 54 \text{ in. } [1.37 \text{ m}]$$

or

$$b_f = \text{center-to-center beam spacing} = 9 \times 12 = 108 \text{ in. } [2.74 \text{ m}]$$

or

$$b_f = \text{beam stem width plus 16 times the slab thickness}$$
$$= 15 + (16 \times 4) = 79 \text{ in. } [2.01 \text{ m}]$$

The limiting value is, therefore, 54 in. [1.37 m]. Next, find the required steel area

$$M_u = 1.2(125) + 1.6(100) = 310 \text{ kip-ft } [420 \text{ kN-m}]$$

$$M_r = \frac{M_u}{0.9} = 344 \text{ kip-ft } [466 \text{ kN-m}]$$

$$A_s = \frac{M_r}{f_y\left(d - \dfrac{t}{2}\right)} = \frac{344 \times 12}{60\left(22 - \dfrac{4}{2}\right)} = 3.44 \text{ in.}^2 \text{ } [2219 \text{ mm}^2]$$

Select bars using Table 6.4, which incorporates consideration for the adequacy of the stem width. From the table choose four No. 9 bars, actual

TABLE 6.4 Options for the T-Beam Reinforcement

Bar Size	No. of Bars	Actual Area Provided (in.²)	Width Required[a] (in.)
7	6	3.60	14
8	5	3.95	13
9	4	4.00	12
10	3	3.81	11
11	3	4.68	11

[a] From Table 9.1.

$A_s = 4.00$ in.². From Table 9.1, the required width for four No. 9 bars is 12 in., less than the 15 in. provided.

Check the concrete stress:

$$C = \frac{M_r}{jd} = \frac{344 \times 12}{20} = 206.4 \text{ kips [918 kN]}$$

$$f_c = \frac{C}{b_f t} = \frac{206.4}{54 \times 4} = 0.956 \text{ ksi [6.59 MPa]}$$

Compare this to the limiting stress of

$$0.85 \, f'_c = 0.85(4) = 3.4 \text{ ksi [23.4 MPa]}$$

Thus, compressive stress in the flange is clearly not critical.

Using the beam stem width of 15 in. and the effective flange width of 54 in., the minimum area of reinforcement is determined as the greater of

$$A_s = \frac{6\sqrt{f'_c}}{F_y}(b_w d) = \frac{6\sqrt{4000}}{60,000}(15 \times 22) = 2.09 \text{ in.}^2 \text{ [1350 mm}^2\text{]}$$

or

$$A_s = \frac{3\sqrt{f'_c}}{F_y}(b_f d) = \frac{3\sqrt{4000}}{60,000} 54 (22) = 3.76 \text{ in.}^2 \text{ [2430 mm}^2\text{]}$$

The minimum area required is, thus, greater than the computed area of 3.44 in.².

The choice of using 4 No. 9 bars is not affected by this development because its area is 4.00 in.². Some of the other data in Table 6.4 must be adjusted if other bar choices are considered.

110   REINFORCED CONCRETE FLEXURAL MEMBERS

The examples in this section illustrate procedures that are reasonably adequate for beams that occur in ordinary slab and beam construction. When special T-sections occur with thin flanges (t less than $d/8$ or so), these methods may not be valid. In such cases, more accurate investigation should be performed, using the requirements of the ACI Code.

Problem 6.5.A. Find the area of steel reinforcement required for a concrete T-beam for the following data: $f'_c = 3$ ksi [20.7 MPa], $f_y = 50$ ksi [345 MPa], $d = 28$ in. [711 mm], $t = 6$ in. [152 mm], $b_w = 16$ in. [406 mm], $b_f = 60$ in. [1520 mm], and the section sustains a factored bending moment of $M_u = 360$ kip-ft [488 kN-m].

Problem 6.5.B. Same as Problem 6.5.A, except $f'_c = 4$ ksi [27.6 MPa], $f_y = 60$ ksi [414 MPa], $d = 32$ in. [813 mm], $t = 5$ in. [127 mm], $b_w = 18$ in. [457 mm], $b_f = 54$ in. [1370 mm], $M_u = 500$ kip-ft [678 kN-m].

6.6 BEAMS WITH COMPRESSION REINFORCEMENT

There are many situations in which steel reinforcement is used on both sides of the neutral axis in a beam. When this occurs, the steel on one side of the axis will be in tension and that on the other side in compression. Such a beam is referred to as a *double reinforced beam* or simply as a *beam with compressive reinforcement* (it being naturally assumed that there is also tensile reinforcement). Various situations involving such reinforcement have been discussed in the preceding sections. In summary, the most common occasions for such reinforcement include:

1. The desired resisting moment for the beam exceeds that for which the concrete alone is capable of developing the necessary compressive force.
2. Other functions of the section require the use of reinforcement on both sides of the beam. These include the need for bars to support U-stirrups and situations when torsion is a major concern.
3. It is desired to reduce deflections by increasing the stiffness of the compressive side of the beam. This is most significant for reduction of long-term creep deflections.
4. The combination of loading conditions on the structure result in reversal moments on the section at a single location; that is, the section must sometimes resist positive moment and at other times resist negative moment.

5. Anchorage requirements (for development of reinforcement) require that the bottom bars in a beam be extended a significant distance into the supports.

The precise investigation and accurate design of doubly reinforced sections, whether performed by the working stress or by strength design methods, are quite complex and are beyond the scope of work in this book. The following discussion presents an approximation method that is adequate for preliminary design of a doubly reinforced section. For real design situations, this method may be used to establish a first trial design, which may then be more precisely investigated using more rigorous methods.

For the beam with double reinforcement, as shown in Figure 6.9a, consider the total resisting moment for the section to be the sum of the following two component moments:

M_1 (Figure 6.9b) is composed of a section with tension reinforcement only (A_{s1}). This section is subject to the usual procedures for design and investigation, as discussed in Section 6.3.

M_2 (Figure 6.9c) is composed of two opposed steel areas (A_{s2} and A'_s) that function in simple moment couple action, similar to the flanges of a steel beam or the top and bottom chords of a truss.

The limit for M_1 is the so-called "balanced moment," as described in Section 6.3. Given the values for steel yield stress and the specified strength of the concrete, the factors for definition of the properties of this balanced section can be obtained from Table 6.1. Given the dimensions for the concrete section, b and d, the limiting moment resistance for the section can be determined using the value of R from the table. If the capacity of the section, as thus determined, is less than the required factored moment (M_r), compressive reinforcing is required. If the balanced resistance is larger than the required factored moment, compressive reinforcement is not required. Although not actually required, the compressive reinforcement may be used for any of the reasons previously mentioned.

When M_1 is less than the factored required moment, it may be determined using the balanced value for R in Table 6.1. Then the required value for M_2 may be determined as $M_2 = M_r - M_1$. This is seldom the case in design work because the amount of reinforcement required to achieve the balanced moment capacity of the section is usually not practical to be fit into the section.

When the potential balanced value for M_1 is greater than the required factored moment, the properties given in Table 6.1 may be used for an ap-

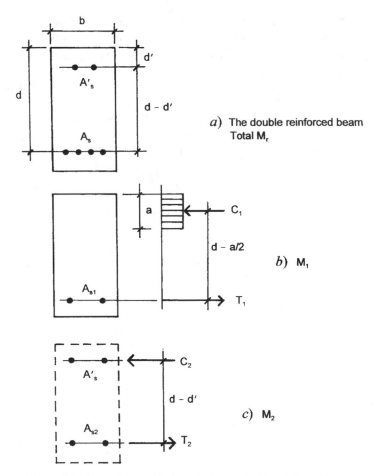

a) The double reinforced beam Total M_r

b) M_1

c) M_2

Figure 6.9 Basis for simplified analysis of a double reinforced beam.

proximation of the value for the required steel area defined as A_{sI}. For the true value of the required M_I, the table values may be adjusted as follows:

1. The actual value for a/d will be smaller than the table value.
2. The actual value for p will be less than the table value.

The approximation procedure that follows starts with the assumption that the concrete dimensions describe a section with the potential M_r

greater than that required. The first step of the procedure is, therefore, to establish what is actually an arbitrary amount of compressive reinforcement. Using this reinforcement as A_{s2}, a value for M_2 is determined. The value for M_1 is then found by subtracting M_2 from the required moment for the section.

Ordinarily, we expect that $A_{s2} = A'_s$ because the same grade of steel is usually used for both. However, there are two special considerations that must be made. The first involves the fact that A_{s2} is in tension, while A'_s is in compression. A'_s must, therefore, be dealt with in a manner similar to that for column reinforcement. This requires, among other things, that the compressive reinforcement be braced against buckling, using ties similar to those in a tied column.

The second consideration for A'_s involves the distribution of stress and strain on the section. Referring to Figure 6.2a, it may be observed that, under normal circumstances (a less than $0.5d$), A'_s will be closer to the neutral axis than A_{s2}. Thus, the stress in A'_s will be lower than that in A_{s2} if pure elastic conditions are assumed. It is common practice to assume steel to be doubly stiff when sharing stress with concrete in compression, due to shrinkage and creep effects. Thus, in translating from linear strain conditions to stress distribution, use the relation $f_c = f'_s/2n$ (where $n = E_s/E_c$).

For the approximate method, it is really not necessary to find separate values for A_{s1} and A_{s2}. This is because of an additional assumption that the value for a is $2 \times d'$. Thus, the moment arm for both A_{s1} and A_{s2} is the same and the value for the total tension reinforcement can be simply determined as

$$A_s = \frac{\text{Required } M_r}{f_y \times (d - d')}$$

This value for A_s can actually be determined as soon as the values for d and d' are established.

With the total tension reinforcement established, the next step involves the determination of A'_s and A_{s2}. Compression reinforcement in beams ordinarily ranges from 0.2 to 0.4 times the total tension reinforcement. For this approximation method, we will determine the area to be $0.3 \times A_s$. We will also assume the stress in the compression reinforcement to be one-half of the yield stress. This permits a definition of the resisting moment M_2 as follows:

$$M_2 = A'_s \left(\frac{f_y}{2} \right)(d - d') = 0.3A_s \left(\frac{f_y}{2} \right)(d - d') = 0.15A_s f_y(d - d')$$

Using this moment, the amount of the tension reinforcement that is required for the development of the steel force couple is one-half of A'_s, and the amount of the tension steel devoted to development of M_1 is found by subtracting this from the total tension reinforcement.

For a final step, the value of A_{s1} may be used to compute a value for the percentage of steel relating to the moment resistance of the section without the compression reinforcement. If this percentage is less than that listed in Table 6.1, the concrete stress will not be critical. However, this situation is predetermined if the total potential balanced resisting moment is determined first and compared to the required factored resisting moment.

For the section that is larger than the balanced section defined by Table 6.1, the procedure can be shortened. The following example will serve to illustrate this procedure:

Example 4. A concrete section with $b = 18$ in. [457 mm] and $d = 21.5$ in. [546 mm] is required to resist service load moments as follows: dead load moment = 175 kip-ft [237 kN-m], live load moment = 160 kip-ft [217 kN-m]. Using strength methods, find the required reinforcement. Use $f'_c = 3$ ksi [20.7 MPa], and $f_y = 60$ ksi [414 MPa].

Solution: Using Table 6.1, find $a/d = 0.592, p = 0.0252$, and $R = 1.063$ in kip-in. units [7330 in kN-m units].

The factored moment for the section is

$$M_u = 1.2(175) + 1.6(160) = 466 \text{ kip-ft [632 kN-m]}$$

and the required factored resisting moment is

$$M_r = \frac{M_u}{0.9} = \frac{466}{0.9} = 518 \text{ kip-ft [702 kN-m]}$$

Using the R value for the balanced section, the maximum resisting moment of the section is

$$M_B = Rbd^2 = \frac{1.063}{12}(18)(21.5)^2 = 737 \text{ kip-ft [999 kN-m]}$$

As this is considerable larger than the required resisting moment, the section is qualified as "underbalanced"; that is, it will be understressed as relates to the compression resistance of the section. It is reasonable, therefore, to use the simplified formula for the tension reinforcing, thus,

$$A_s = \frac{\text{Required } M_r}{f_y \times (d - d')} = \frac{518 \times 12}{60(21.5 - 2.5)} = 5.45 \text{ in.}^2 \, [3520 \text{ mm}^2]$$

and a reasonable assumption for the compressive reinforcement is

$$A'_s = 0.3A_s = 0.3(5.45) = 1.63 \text{ in.}^2 \, [1050 \text{ mm}^2]$$

Bar combinations may next be found for these two steel areas. If you wish to use a given concrete section for a resisting moment that exceeds the balanced limit described by the values in Table 6.1, a different procedure is required. In this case, the first two steps are the same as in the preceding example: determining the required resisting moment M_r and the limiting balanced moment M_B. The tension reinforcement required for the balanced moment can be determined with the balanced percentage p from Table 6.1. M_B then becomes the moment M_1, as shown in Figure 6.9b. M_2, as shown in Figure 6.9c, is the difference between M_r and M_B. The compression reinforcement and the additional tension reinforcement required for M_2 can then be determined. The following example illustrates this procedure:

Example 5. Find the reinforcement required for the beam in Example 4 if the required resisting moment M_r is 900 kip-ft [1220 kN-m].

Solution: The first step is the determination of the limiting balanced moment M_B. For this section, this value was computed in Example 4 as 737 kip-ft. As the required moment exceeds this value, compression reinforcement is required, and the moment for this is determined as

$$M_2 = M_r - M_B = 900 - 737 = 163 \text{ kip-ft} \, [221 \text{ kN-m}]$$

For the balanced moment, the required tension reinforcement can be computed using the balanced p from Table 6.1. Thus,

$$A_{s1} = pbd = 0.0252 \times 18 \times 21.5 = 9.75 \text{ in.}^2 \, [6290 \text{ mm}^2]$$

For the determination of the tension reinforcement required for M_2, the procedure involves the use of the moment arm $d - d'$. Thus,

$$A_{s2} = \frac{M_2}{f_y(d - d')} = \frac{163 \times 12}{60(19)} = 1.72 \text{ in.}^2 \text{ [1110 mm}^2]$$

and the total required tension reinforcing is

$$A_s = 9.75 + 1.72 = 11.5 \text{ in.}^2 \text{ [7420 mm}^2]$$

For the compressive reinforcement, we assume a stress approximately equal to one-half the yield stress. Thus, the area of steel required is twice the value of A_{s2}, $2(1.72) = 3.44 \text{ in.}^2$ [2220 mm^2]. This requirement can be met with 3 No. 10 bars providing an area of 3.81 in.^{2} [2460 mm^2].

Options for the tension reinforcement are given in Table 6.5. As the table indicates, it is not possible to get the bars into the 18 in.-wide beam by placing them in a single layer. Options are to use two layers of bars or to increase the beam width. Frankly, this is not a good beam design and would most likely not be acceptable unless extreme circumstances force the use of the limited beam size. This is pretty much the typical situation for beams required to develop resisting moments larger than the balanced moment.

Problem 6.6.A. A concrete section with $b = 16$ in. [406 mm] and $d = 19.5$ in. [495 mm] is required to develop a bending moment strength of 400 kip-ft [542 kN-m]. Use of compressive reinforcement is desired. Find the required reinforcement. Use $f'_c = 4$ ksi [27.6 MPa] and $f_y = 60$ ksi [414 MPa].

Problem 6.6.B. Same as Problem 6.6.A, except required $M_r = 1000$ kip-ft [1360 kN-m], $b = 20$ in. [508 mm], $d = 27$ in. [686 mm].

TABLE 6.5 Options for the Tension Reinforcement—Example 5

Bar Size	Area of One Bar (in.^{2})	No. of Bars	Actual Area Provided (in.^{2})	Width Required[a] (in.)
8	0.79	15	11.85	33
9	1.00	12	12.00	30
10	1.27	10	12.70	28
11	1.56	8	12.48	25

[a] See Section 9.1.

Problem 6.6.C.   Same as Problem 6.6.A, except M_r =640 kip-ft [868 kN-m].

Problem 6.6.D.   Same as Problem 6.6.B, except M_r = 1400 kip-ft [1900 kN-m].

6.7   SPANNING SLABS

Concrete slabs are frequently used as spanning roof or floor decks, often occurring in monolithic, cast-in-place slab and beam framing systems. There are generally two basic types of slabs: one-way spanning and two-way spanning. The spanning condition is not determined so much by the slab as by its support conditions. As part of a general framing system, the one-way spanning slab is discussed in Section 9.1. The following discussion relates to the design of one-way solid slabs using procedures developed for the design of rectangular beams.

Solid slabs are usually designed by considering the slab to consist of a series of 12-in.-wide planks. Thus, the procedure consists of simply designing a beam section with a predetermined width of 12 in. Once the depth of the slab is established, the required area of steel is determined, specified as the number of square inches of steel required per foot of slab width.

Reinforcing bars are selected from a limited range of sizes, appropriate to the slab thickness. For thin slabs (4 to 6 in. thick), bars may be of a size from No. 3 to No. 6 or so (nominal diameters from 3/8 to 3/4 in.). The bar size selection is related to the bar spacing, the combination resulting in the amount of reinforcing in terms of so many square inches per one foot unit of slab width. Spacing is limited by code regulation to a maximum of three times the slab thickness. There is no minimum spacing, other than that required for proper placing of the concrete; however, a very close spacing indicates a very large number of bars, making for laborious installation.

Every slab must be provided with two-way reinforcement, regardless of its structural functions. This is partly to satisfy requirements for shrinkage and temperature effects. The amount of this minimum reinforcement is specified as a percentage p of the gross cross-sectional area of the concrete as follows:

1. For slabs reinforced with grade 40 or grade 50 bars:

$$p = \frac{A_s}{bt} = 0.002 \quad (0.2\%)$$

2. For slabs reinforced with grade 60 bars:

$$p = \frac{A_s}{bt} = 0.0018 \quad (0.18\%)$$

Center-to-center spacing of this minimum reinforcement must not be greater than five times the slab thickness or 18 in.

Minimum cover for slab reinforcement is normally 0.75 in., although exposure conditions or need for a high fire rating may require additional cover. For a thin slab reinforced with large bars, there will be a considerable difference between the slab thickness t and the effective depth d, as shown in Figure 6.10. Thus, the practical efficiency of the slab in flexural resistance decreases rapidly as the slab thickness is decreased. For this and other reasons, very thin slabs (less than 4 in. thick) are often reinforced with wire fabric rather than sets of loose bars.

Shear reinforcement is seldom used in one-way slabs, and consequently the maximum unit shear stress in the concrete must be kept within the limit for the concrete without reinforcement. This is usually not a concern because unit shear is usually low in one-way slabs, except for exceptionally high loadings.

Table 6.6 gives data that are useful in slab design, as demonstrated in the following example. Table values indicate the average amount of steel area per foot of slab width provided by various combinations of bar size and spacing. Table entries are determined as follows:

$$A_s/\text{ft} = (\text{single bar area}) \frac{12}{\text{bar spacing}}$$

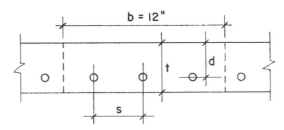

Figure 6.10 Reference for slab design.

TABLE 6.6 Areas Provided by Spaced Reinforcement

Bar Spacing (in.)	Area Provided (in.²/ft width)									
	No. 2	No. 3	No. 4	No. 5	No. 6	No. 7	No. 8	No. 9	No. 10	No. 11
3	0.20	0.44	0.80	1.24	1.76	2.40	3.16	4.00		
3.5	0.17	0.38	0.69	1.06	1.51	2.06	2.71	3.43	4.35	
4	0.15	0.33	0.60	0.93	1.32	1.80	2.37	3.00	3.81	4.68
4.5	0.13	0.29	0.53	0.83	1.17	1.60	2.11	2.67	3.39	4.16
5	0.12	0.26	0.48	0.74	1.06	1.44	1.89	2.40	3.05	3.74
5.5	0.11	0.24	0.44	0.68	0.96	1.31	1.72	2.18	2.77	3.40
6	0.10	0.22	0.40	0.62	0.88	1.20	1.58	2.00	2.54	3.12
7	0.08	0.19	0.34	0.53	0.75	1.03	1.35	1.71	2.18	2.67
8	0.07	0.16	0.30	0.46	0.66	0.90	1.18	1.50	1.90	2.34
9	0.07	0.15	0.27	0.41	0.59	0.80	1.05	1.33	1.69	2.08
10	0.06	0.13	0.24	0.37	0.53	0.72	0.95	1.20	1.52	1.87
11	0.05	0.12	0.22	0.34	0.48	0.65	0.86	1.09	1.38	1.70
12	0.05	0.11	0.20	0.31	0.44	0.60	0.79	1.00	1.27	1.56
13	0.05	0.10	0.18	0.29	0.40	0.55	0.73	0.92	1.17	1.44
14	0.04	0.09	0.17	0.27	0.38	0.51	0.68	0.86	1.09	1.34
15	0.04	0.09	0.16	0.25	0.35	0.48	0.63	0.80	1.01	1.25
16	0.04	0.08	0.15	0.23	0.33	0.45	0.59	0.75	0.95	1.17
18	0.03	0.07	0.13	0.21	0.29	0.40	0.53	0.67	0.85	1.04
24	0.02	0.05	0.10	0.15	0.22	0.30	0.39	0.50	0.63	0.78

Thus, for No. 5 bars at 8-in. centers,

$$A_s/\text{ft} = (0.31)\left(\frac{12}{8}\right) = 0.465 \text{ in.}^2/\text{ft}$$

It may be observed that the table entry for this combination is rounded off to a value of 0.46 in.²/ft.

Example 6. A one-way, solid concrete slab is to be used for a simple span of 14 ft [4.27 m]. In addition to its own weight, the slab carries a superimposed dead load of 30 psf [1.44 kPa] plus a live load of 100 psf [4.79 kPa]. Using $f'_c = 3$ ksi [20.7 MPa] and $f_y = 40$ ksi [276 MPa], design the slab for minimum overall thickness.

Solution: Using the general procedure for design of a beam with a rectangular section (Section 6.2), we first determine the required slab thickness. Thus, for deflection, from Table 6.8,

$$\text{Minimum } t = \frac{L}{25} = \frac{14 \times 12}{25} = 6.72 \text{ in. [171 mm]}$$

For flexure, first determine the maximum bending moment. The loading must include the weight of the slab, for which the thickness required for deflection may be used as a first estimate. Assuming a 7-in. [178 mm]-thick slab, then slab weight is $(7/12)(150 \text{ pcf}) = 87.5 \text{ psf}$, say 88 psf, and the total dead load is $30 + 88 = 118 \text{ psf [5.65 kPa]}$.

The factored load is, thus,

$$U = 1.2(\text{dead load}) + 1.6(\text{live load}) = 1.2(118) + 1.6(100)$$

$$= 302 \text{ psf [14.45 kPa]}$$

The maximum bending moment on a 12-in.-wide strip of the slab, thus, becomes

$$M_u = \frac{wL^2}{8} = \frac{302(14)^2}{8} = 7399 \text{ ft-lb [10.0 kN-m]}$$

and the required factored resisting moment is

$$M_r = \frac{7399}{0.9} = 8221 \text{ ft-lb [11.2 kN-m]}$$

For a minimum slab thickness, we consider the use of a balanced section, for which Table 6.1 yields the following properties: $a/d = 0.685$, and $R = 1.149$ (in kip and inch units).

Then the minimum value for bd^2 is

$$bd^2 = \frac{M_r}{R} = \frac{8.221 \times 12}{1.149} = 85.9 \text{ in.}^3 \text{ [1,400,00 mm}^3\text{]}$$

and since b is the 12-in. design strip width,

$$d = \sqrt{\frac{85.9}{12}} = \sqrt{7.16} = 2.68 \text{ in. [68 mm]}$$

Assuming an average bar size of a No. 6 (¾ in. nominal diameter) and cover of ¾ in., the minimum required slab thickness based on flexure becomes

$$t = d + 2(\text{bar diameter}) + \text{cover}$$

$$t = 2.68 + \frac{0.75}{2} + 0.75 = 3.8 \text{ in. } [96.5 \text{ mm}]$$

The deflection limitation thus controls in this situation, and the minimum overall thickness is the 6.72-in. dimension. Staying with the 7-in overall thickness, the actual effective depth with a No. 6 bar will be

$$d = 7.0 - 1.125 = 5.875 \text{ in. } [149 \text{ mm}]$$

Since this d is larger than that required for a balanced section, the value for a/d will be slightly smaller than 0.685, as found from Table 6.1. Assume a value of 0.4 for a/d, and determine the required area of reinforcement as follows:

$$a = 0.4d = 0.4(5.875) = 2.35 \text{ in. } [59.7 \text{ mm}]$$

$$A_s = \frac{M}{f_y\left(d - \frac{a}{2}\right)} = \frac{8.221 \times 12}{40(5.875 - 1.175)} = 0.525 \text{ in.}^2 \ [339 \text{ mm}^2]$$

Using data from Table 6.6, the optional bar combinations shown in Table 6.7 will satisfy this requirement. Note that for bars larger than the assumed No. 6 bar (0.75 in. diameter), d will be slightly less and the required area of reinforcement slightly higher.

The ACI Code permits a maximum center-to-center bar spacing of three times the slab thickness (21 in. in this case) or 18 in., whichever is smaller. Minimum spacing is largely a matter of the designer's judgment. Many designers consider a minimum practical spacing to be one approximately equal to the slab thickness. Within these limits, any of the bar size and spacing combinations listed are adequate.

TABLE 6.7   Alternatives for the Slab Reinforcement

Bar Size	Spacing of Bars Center to Center (in.)	Average A_s in a 12-in. Width (in.²)
5	7	0.53
6	10	0.53
7	13	0.55
8	18	0.53

As described previously, the ACI Code requires a minimum reinforcement for shrinkage and temperature effects to be placed in the direction perpendicular to the flexural reinforcement. With the grade 40 bars in this example, the minimum percentage of this steel is 0.0020, and the steel area required for a 12-in. strip, thus, becomes

$$A_s = p(bt) = 0.0020(12 \times 7) = 0.168 \text{ in.}^2/\text{ft} \ [356 \text{ mm}^2]$$

From Table 6.6, this requirement can be satisfied with No. 3 bars at 7-in. centers or No. 4 bars at 14-in. centers. Both of these spacings are well below the maximum of five times the slab thickness (35 in.) or 18 in.

Although simply supported single slabs are sometimes encountered, the majority of slabs used in building construction are continuous through multiple spans. An example of the design of such a slab is given in Chapter 9.

Problem 6.7.A.   A one-way solid concrete slab is to be used for a simple span of 16 ft [4.88 m]. In addition to its own weight, the slab carries a superimposed dead load of 40 psf [1.92 kPa] and a live load of 100 psf [4.79 kPa]. Using the strength method with $f'_c = 3$ ksi [20.7 MPa] and $f_y = 40$ ksi [276 MPa], design the slab for minimum overall thickness.

Problem 6.7.B.   Same as Problem 6.7.A, except span = 18 ft [5.49 m], superimposed dead load = 50 psf [2.39 kPa], live load = 75 psf [3.59 kPa], $f'_c = 4$ ksi [27.6 MPa], $f_y = 60$ ksi [414 MPa].

6.8 DEFLECTION CONTROL

Deflection of spanning slabs and beams of cast-in-place concrete is controlled primarily by using a recommended minimum thickness (overall height) expressed as a percentage of the span. Table 6.8 is adapted from a similar table given in the ACI Code and yields minimum thickness as a fraction of the span. Table values apply only for concrete of normal weight (made with ordinary sand and gravel) and for reinforcement with f_y of 40 ksi [276 MPa] and 60 ksi [414 MPa]. The ACI Code supplies correction factors for other concrete weights and reinforcing grades. The ACI Code further stipulates that these recommendations apply only where beam deflections are not critical for other elements of the building construction, such as supported partitions subject to cracking caused by beam deflections.

TABLE 6.8 Minimum Thickness of Slabs or Beams Unless Deflections Are Computed[a]

Type of Member	End Conditions of Span	Minimum Thickness of Slab or Height of Beam	
		$f_y = 40$ ksi [276 MPa]	$f_y = 60$ ksi [414 MPa]
Solid One-Way Slabs	Simple support	$L/25$	$L/20$
	One end continuous	$L/30$	$L/24$
	Both ends continuous	$L/35$	$L/28$
	Cantilever	$L/12.5$	$L/10$
Beams or Joists	Simple aupport	$L/20$	$L/16$
	One end continuous	$L/23$	$L/18.5$
	Both ends continuous	$L/26$	$L/21$
	Cantilever	$L/10$	$L/8$

[a]Refers to overall vertical dimension of concrete section. For normal weight concrete (145 pcf) only; code provides adjustment for other weights. Valid only for members not supporting or attached rigidly to partitions or other construction likely to be damaged by large deflections.

Source: Adapted from material in *Building Code Requirements for Structural Concrete (ACI 318–02)*, with permission of the publishers, American Concrete Institute.

Deflection of concrete structures presents a number of special problems. For concrete with ordinary reinforcement (not prestressed), flexural action normally results in some tension cracking of the concrete at points of maximum bending. Thus, the presence of cracks in the bottom of a beam at midspan points and in the top over supports is to be expected. In general, the size (and visibility) of these cracks will be proportional to the amount of beam curvature produced by deflection. Crack size will also be greater for long spans and for deep beams. If visible cracking is considered objectionable, more conservative depth-to-span ratios should be used, especially for spans over 30 ft and beam depths over 30 in.

Creep of concrete results in additional deflections over time. This is caused by the sustained loads—essentially the dead load of the construction. Deflection controls reflect concern for this, as well as for the instantaneous deflection under live load, the latter being the major concern in structures of wood and steel.

In beams, deflections, especially creep deflections, may be reduced by the use of some compressive reinforcement. Where deflections are of concern, or where depth-to-span ratios are pushed to their limits, it is advisable to use some compressive reinforcement, consisting of continuous top bars.

When, for whatever reasons, deflections are deemed to be critical, computations of actual values of deflection may be necessary. The ACI Code provides directions for such computations; they are quite complex in most cases, and beyond the scope of this work. In actual design work, however, they are required very infrequently.

7

SHEAR IN CONCRETE STRUCTURES

There are many situations in concrete structures that involve the development of shear. In most cases, the shear stress itself is not the major concern but rather the diagonal tension that accompanies the shear action. In many situations, shear stresses are quite low in magnitude and may well be within the capacity of the concrete without any added reinforcement. In other situations, the reinforcement provided for other purposes may assist in resistance to shear effects. This chapter deals with investigation for various shear actions and with some methods for enhancing the concrete structure with reinforcement when shear is excessive.

7.1 GENERAL CONCERNS FOR SHEAR

The most common situations involving shear in concrete structures are shown in Figure 7.1. Shear in beams (Figure 7.1a) is ordinarily critical near the supports, where the shear force is greatest. In short brackets (Figure 7.1b) and keys (Figure 7.1c), the shear action is essentially a di-

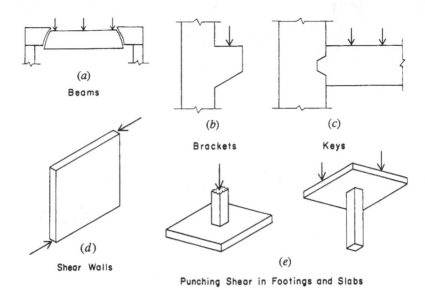

(a)
Beams

(b)
Brackets

(c)
Keys

(d)
Shear Walls

(e)
Punching Shear in Footings and Slabs

Figure 7.1   Situations involving shear in concrete structures.

rect slicing effect. *Punching shear,* also called *peripheral shear,* occurs in column footings and in slabs that are directly supported on columns (Figure 7.1*e*). When walls are used as bracing elements for shear forces that are parallel to the wall surface (called *shear walls*), they must develop resistance to the direct shear effect that is similar to that in a bracket (Figure 7.1*d*). In all of these situations, consideration must be given to the shear effect and the resulting shear stresses. Both the magnitude and the direction of the shear stresses must be considered. In many cases, however, the shear effect occurs in combination with other effects, such as bending moment, axial tension, or axial compression. In combined force situations, the resulting net combined stress situations must be considered.

7.2   SHEAR IN BEAMS

From general consideration of shear effects, as developed in the science of mechanics of materials, the following observations can be made:

1. Shear is an ever-present phenomenon, produced directly by slicing actions, by lateral loading in beams, and on oblique sections in tension and compression members.

2. Shear forces produce shear stress in the plane of the force and equal unit shear stresses in planes that are perpendicular to the shear force.

3. Diagonal stresses of tension and compression, having magnitudes equal to that of the shear stress, are produced in directions of 45° from the plane of the shear force.

4. Direct slicing shear force produces a constant magnitude shear stress on affected sections, but beam shear action produces shear stress that varies on the affected sections, having magnitude of zero at the edges of the section and a maximum value at the centroidal neutral axis of the section.

In the discussions that follow it is assumed that the reader has a general familiarity with these relationships.

Consider the case of a simple beam with uniformly distributed load and end supports that provide only vertical resistance (no moment re straint). The distribution of internal shear and bending moment are shown in Figure 7.2a.

For flexural resistance, it is necessary to provide longitudinal reinforcing bars near the bottom of the beam. These bars are oriented for primary effectiveness in resistance to tension stresses that develop on a vertical (90°) plane (which is the case at the center of the span, where the bending moment is maximum and the shear approaches zero).

Under the combined effects of shear and bending, the beam tends to develop tension cracks as shown in Figure 7.2b. Near the center of the span, where the bending is predominant and the shear approaches zero, these cracks approach 90°. Near the support, however, where the shear predominates and bending approaches zero, the critical tension stress plane approaches 45°, and the horizontal bars are only partly effective in resisting the cracking.

7.3 SHEAR REINFORCEMENT FOR BEAMS

For beams, the most common form of added shear reinforcement consists of a series of U-shaped bent bars (Figure 7.2d), placed vertically and spaced along the beam span, as shown in Figure 7.2c. These bars, called *stirrups*, are intended to provide a vertical component of resistance, working in conjunction with the horizontal resistance provided by the flexural reinforcement. In order to develop flexural tension near the support face, the horizontal bars must be bonded to the concrete beyond the

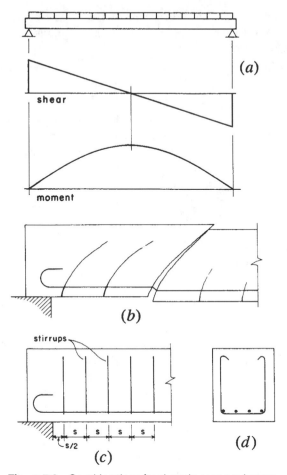

Figure 7.2   Considerations for shear in concrete beams.

point where the stress is developed. Where the ends of simple beams extend only a short distance over the support (a common situation), it is often necessary to bend or hook the bars, as shown in Figure 7.2c.

The simple span beam and the rectangular section shown in Figure 7.2d occur only infrequently in building structures. The most common case is that of the beam section shown in Figure 7.3, which occurs when a beam is cast continuously with a supported concrete slab. In addition, these beams normally occur in continuous spans with negative moments

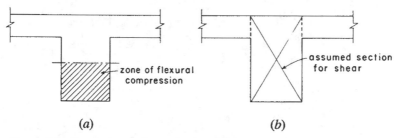

Figure 7.3   Development of negative bending and shear in concrete T-beams.

at the supports. Thus, the stress in the beam near the support is as shown
in Figure 7.3a, with the negative moment producing compressive flexural
stress in the bottom of the beam stem. This is substantially different from
the case of the simple beam, where the moment approaches zero near the
support.

For the purpose of shear resistance, the continuous, T-shaped beam is
considered to consist of the section indicated in Figure 7.3b. The effect
of the slab is ignored, and the section is considered to be a simple rectangular one. Thus, for shear design, there is little difference between the
simple span beam and the continuous beam, except for the effect of the
continuity on the distribution of shear along the beam span. It is important, however, to understand the relationships between shear and moment
in the continuous beam.

Figure 7.4 illustrates the typical condition for an interior span of a continuous beam with uniformly distributed load. Referring to the portions
of the beam span numbered 1, 2, and 3:

1. In this zone, the high negative moment requires major flexural reinforcement consisting of horizontal bars near the top of the beam.

2. In this zone, the moment reverses sign; moment magnitudes are
   low; and, if shear stress is high, the design for shear is a predominant concern.

3. In this zone, shear consideration is minor, and the predominant
   concern is for positive moment requiring major flexural reinforcement in the bottom of the beam.

Vertical U-shaped stirrups, similar to those shown in Figure 7.5a, may
be used in the T-shaped beam. An alternate detail for the U-shaped stirrup is shown in Figure 7.5b, in which the top hooks are turned outward;

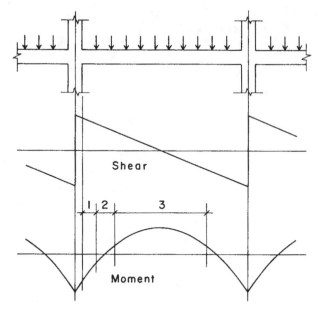

Figure 7.4 Shear and bending in continuous beams.

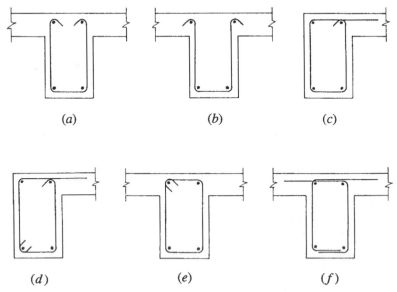

Figure 7.5 Forms for vertical stirrups.

this makes it possible to spread the negative moment reinforcing bars to make placing of the concrete somewhat easier. Figures 7.5c and d show possibilities for stirrups in L-shaped beams that occur at the edges of large openings or at the outside edge of the structure. This form of stirrup is used to enhance the torsional resistance of the section and also assists in developing the negative moment resistance in the slab at the edge of the beam.

So-called *closed stirrups*, similar to ties in columns, are sometimes used for T- and L-shaped beams, as shown in Figures 7.5c through f. These are generally used to improve the torsional resistance of the beam section.

Stirrup forms are often modified by designers or by the reinforcing fabricator's detailers to simplify the fabrication and/or the field installation. The stirrups shown in Figures 7.5d and f are two such modifications of the basic details in Figures 7.5c and e, respectively.

The following are considerations and code requirements that apply to design for beam shear:

Concrete Capacity. Whereas the tensile strength of the concrete is ignored in design for flexure, the concrete is assumed to take some portion of the shear in beams. If the capacity of the concrete is not exceeded— as is sometimes the case for lightly loaded beams—there may be no need for reinforcement. The typical case, however, is shown in Figure 7.6, where the maximum shear V exceeds the capacity of the concrete alone (V_c) and the steel reinforcement is required to absorb the excess, indicated as the shaded portion in the shear diagram.

Minimum Shear Reinforcement. Even when the maximum computed shear stress falls below the capacity of the concrete, the present code requires the use of some minimum amount of shear reinforcement. Exceptions are made in some situations, such as for slabs and very shal-

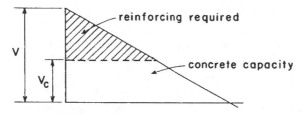

Figure 7.6 Sharing of shear resistance in reinforced concrete beams.

low beams. The objective is essentially to toughen the structure with a small investment in additional reinforcement.

Type of Stirrup. The most common stirrups are the simple U-shape or closed forms shown in Figure 7.5, placed in a vertical position at intervals along the beam. It is also possible to place stirrups at an incline (usually 45°), which makes them somewhat more effective in direct resistance to the potential shear cracking near the beam ends (see Figure 7.2). In large beams with high unit shear stress, both vertical and inclined stirrups are sometimes used at the location of the greatest shear.

Size of Stirrups. For beams of moderate size, the most common size for U-stirrups is a No. 3 bar. These bars can be bent relatively tightly at the corners (small radius of bend) in order to fit within the beam section. For larger beams, a No. 4 bar is sometimes used, its strength (as a function of its cross-sectional area) being almost twice that of a No. 3 bar.

Spacing of Stirrups. Stirrup spacings are computed (as discussed in the following sections) on the basis of the amount of reinforcing required for the unit shear stress at the location of the stirrups. A maximum spacing of $d/2$ (i.e., one-half the effective beam depth d) is specified in order to ensure that at least one stirrup occurs at the location of any potential diagonal crack (see Figure 7.2). When shear stress is excessive, the maximum spacing is limited to $d/4$.

Critical Maximum Design Shear. Although the actual maximum shear value occurs at the end of the beam, the code permits the use of the shear stress at a distance of d (effective beam depth) from the beam end as the critical maximum for stirrup design. Thus, as shown in Figure 7.7, the shear requiring reinforcement is slightly different from that shown in Figure 7.6.

Total Length for Shear Reinforcement. On the basis of computed shear forces, reinforcement must be provided along the beam length for the distance defined by the shaded portion of the shear stress diagram shown in Figure 7.7. For the center portion of the span, the concrete is theoretically capable of the necessary shear resistance without the assistance of reinforcement. However, the code requires that some shear reinforcement be provided for a distance beyond this computed cutoff point. Earlier codes required that stirrups be provided for a distance equal to the

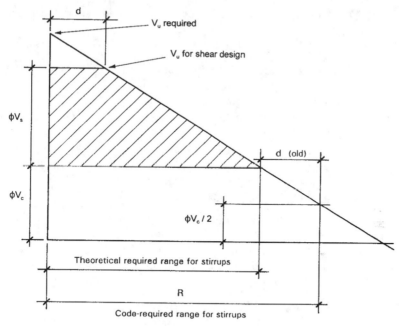

Figure 7.7   Layout for shear stress analysis: ACI Code requirements.

effective depth of the beam beyond the computed cutoff point. Currently, codes require that minimum shear reinforcement be provided as long as the computed shear force exceeds one-half of the capacity of the concrete ($\phi \times V_c/2$). However it is established, the total extended range over which reinforcement must be provided is indicated as R on Figure 7.7.

7.4   DESIGN FOR BEAM SHEAR

The following is a description of the procedure for design of shear reinforcement for beams that are experiencing flexural and shear stresses exclusively.

The ultimate shear force (V_u) at any cross section along a given beam due to factored loading must be less than the reduced shear capacity at the section. Mathematically this is represented as:

$$V_u \le \phi_v \times (V_c + V_s)$$

where V_u = ultimate shear force at the section due to factored loading
   ϕ_v = 0.75
   V_c = shear capacity of the concrete
   V_s = shear capacity of the steel reinforcement

For beams of normal weight concrete, subjected only to flexure and shear, shear force in the concrete is limited to

$$V_c = 2\sqrt{f'_c}\,bd$$

where f'_c = specified strength of the concrete in psi
   b = width of the cross section in in.
   d = effective depth of the cross section in in.

When V_u exceeds the limit for $\phi_v V_c$, reinforcing must be provided, complying with the general requirements discussed previously. Thus,

$$V_s \geq \frac{V_u}{\phi_v} - V_c$$

Required spacing of shear reinforcement is determined as follows. Referring to Figure 7.8, note that the capacity in tensile resistance of a single, two-legged stirrup is equal to the product of the total steel cross-sectional area, A_v, times the yield steel stress. Thus,

$$T = A_v f_y$$

This resisting force opposes part of the steel shear force required at the location of the stirrup, referred to as V_s'. Equating the stirrup tension to this force, an equilibrium equation for one stirrup is obtained; thus,

$$A_v f_y = V_s'$$

The total shear force capacity of the beam in excess of the concrete is determined by the number of stirrups encountered by the shear force acting at a 45° angle through the beam. The number of stirrups will be d/s. Thus, the equilibrium equation for the beam is

$$\left(\frac{d}{s}\right) A_v f_y = V_s'$$

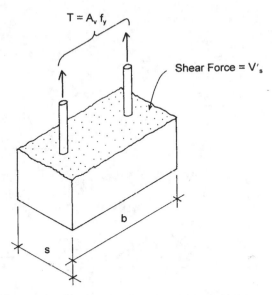

$T = A_v f_y$

Shear Force = V'_s

b

s

Figure 7.8 Consideration for spacing of a single stirrup.

From this equation, an expression for the required spacing can be derived; thus,

$$s \leq \frac{A_v f_y d}{V'_s}$$

The following example illustrates the design procedure for a simple beam:

Example 1. Design the required shear reinforcement for the simple beam shown in Figure 7.9. Use $f'_c = 3$ ksi [20.7 MPa] and $f_y = 40$ ksi [276 MPa] and single U-shaped stirrups.

Solution: First, the loading must be factored in order to determine the ultimate shear force.

$$w_u = 1.2 \times w_{DL} + 1.6 \times w_{LL} = 1.2 \times 2 \text{ klf} + 1.6 \times 3 \text{ klf}$$

$$= 7.2 \text{ klf } [105 \text{ kN/m}]$$

The maximum value for the shear is 57.6 kips [256 kN].

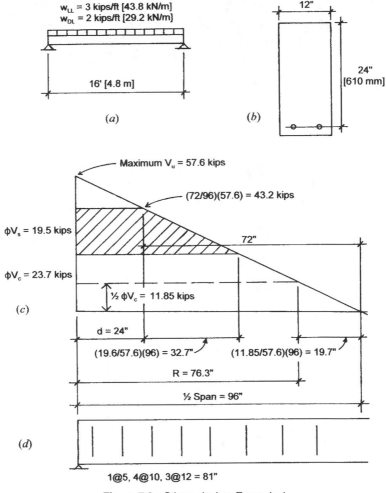

Figure 7.9 Stirrup design: Example 1.

Next, construct the ultimate shear force diagram (V_u) for one-half of the beam, as shown in Figure 7.9c. For the shear design, the critical shear force is at 24 in. (the effective depth of the beam) from the support. Using proportionate triangles, this value is

$$V_u = \left(\frac{72\ \text{in.}}{96\ \text{in.}}\right)(57.6\ \text{kips}) = 43.2\ \text{kips [192 kN]}$$

The shear capacity of the concrete without reinforcing is

$$\phi V_c = (0.75)2\sqrt{f'_c}\,(b)(d)$$

$$= (0.75)2\sqrt{3000}\ \text{psi}(12\ \text{in.})(24\ \text{in.})$$

$$= 23{,}700\ \text{lbs} = 23.7\ \text{kips}\ [105\ \text{kN}]$$

At the point of critical force, therefore, there is an excess shear force of $43.2 - 23.7 = 19.5$ kips [87 kN] that must be carried by reinforcement. Next, complete the construction of the diagram in Figure 7.9c to define the shaded portion, which indicates the extent of the required reinforcement. Observe that the excess shear condition extends to 56.7 in. [1.44 m] from the support.

In order to satisfy the requirements of the ACI Code, shear reinforcement must be used wherever the shear force (V_u) exceeds one-half of ϕV_c. As shown in Figure 7.9c, this is a distance of 76.3 in. from the support. The code further stipulates that the minimum cross-sectional area of this reinforcing be

$$A_v = 50\left[\frac{b \times s_{max}}{f_y}\right]$$

With $f_y = 40$ ksi [276 MPa] and the maximum allowable spacing of one-half the effective depth, the required area is

$$A_v = 50\left[\frac{12\ \text{in.} \times 12\ \text{in.}}{40{,}000\ \text{psi}}\right] = 0.18\ \text{in.}^2\ [116\ \text{mm}^2]$$

which is less than the area of $2 \times 0.11 = 0.22$ in.^{2} provided by the two legs of the No. 3 stirrup.

For the maximum V_s value of 19.5 kips, the maximum spacing permitted at the critical point 24 in. from the support is determined as

$$s \leq \frac{A_v f_y d}{V_s} = \frac{(0.22\ \text{in.}^2)(40\ \text{ksi})(24\ \text{in.})}{19.5\ \text{kips}} = 10.8\ \text{in.}\ [274\ \text{mm}]$$

Since this is less than the maximum allowable of one-half the depth or 12 in., it is best to calculate at least one more spacing at a short distance beyond the critical point. For example, at 36 in. from the support the shear force is

$$V_u = \left(\frac{60 \text{ in.}}{96 \text{ in.}}\right)(57.6 \text{ kips}) = 36.0 \text{ kips } [160 \text{ kN}]$$

and the value of V'_s at this point is $36.0 - 23.7 \text{ kips} = 12.3 \text{ kips}$. The spacing required at this point is, thus,

$$s \leq \frac{A_v f_y d}{V'_s} = \frac{(0.22 \text{ in.}^2)(40 \text{ ksi})(24 \text{ in.})}{12.4 \text{ kips}} = 17.2 \text{ in. } [437 \text{ mm}]$$

which indicates that the required spacing drops to the maximum allowed at less than 12 in. from the critical point. A possible choice for the stirrup spacings is shown in Figure 7.9d, with a total of eight stirrups that extend over a range of 81 in. from the support. There are thus a total of 16 stirrups in the beam, 8 at each end. Note that the first stirrup is placed at 5 in. from the support, which is one-half the computed required spacing; this is a common practice with designers.

Example 2.   Determine the required number and spacings for No. 3 U-stirrups for the beam shown in Figure 7.10. Use $f'_c = 3$ ksi [20.7 MPa] and $f_y = 40$ ksi [276 MPa].

Solution: As in Example 1, the shear values are determined, and the diagram in Figure 7.10c. is constructed. In this case, the maximum critical shear force of 28.5 kips and a shear capacity of concrete (ϕV_c) of 16.4 kips results in a maximum ϕV_s value to 12.1 kips, for which the required spacing is

$$s \leq \frac{A_v f_y d}{V'_s} = \frac{(0.22 \text{ in.}^2)(40 \text{ ksi})(20 \text{ in.})}{12.1 \text{ kips}} = 14.5 \text{ in. } [368 \text{ mm}]$$

Since this value exceeds the maximum limit of $d/2 = 10$ in., the stirrups may all be placed at the limited spacing, and a possible arrangement is as shown in Figure 7.10d. As in Example 1, note that the first stirrup is placed at one-half the required distance from the support.

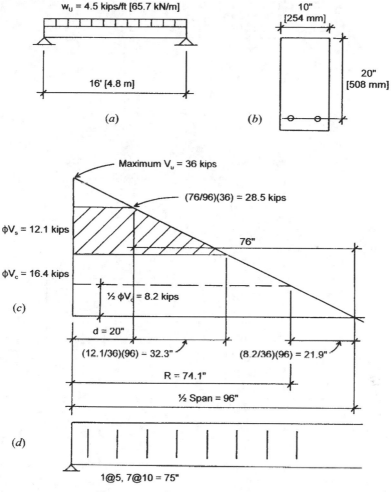

Figure 7.10   Stirrup design: Example 2.

Example 3.   Determine the required number and spacings for No. 3 U-stirrups for the beam shown in Figure 7.11. Use $f'_c = 3$ ksi [20.7 MPa] and $f_y = 40$ ksi [276 MPa].

Solution: In this case, the maximum critical design shear force is found to be less than V_c, which in theory indicates that reinforcement is not required. To comply with the code requirement for minimum reinforce-

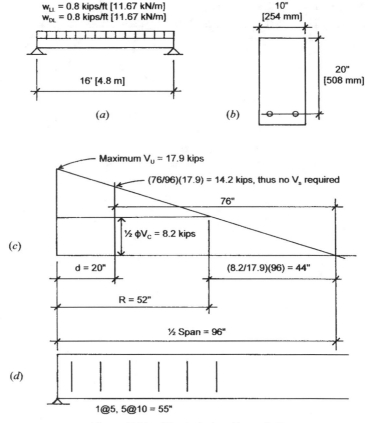

Figure 7.11   Stirrup design: Example 3.

ment, however, provide stirrups at the maximum permitted spacing out to the point where the shear stress drops to 8.2 kips (one-half of ϕV_c). To verify that the No. 3 stirrup is adequate, compute

$$A_v = 50 \left[\frac{10 \text{ in.} \times 10 \text{ in.}}{40,000 \text{ psi}} \right] = 0.125 \text{ in.}^2$$

which is less than the area of 0.22 in. provided, so the No. 3 stirrup at 10-in. is adequate.

Examples 1 through 3 have illustrated what is generally the simplest case for beam shear design - that of a beam with uniformly distributed load and with sections subjected only to flexure and shear. When concentrated loads or unsymmetrical loadings produce other forms for the shear diagram, these must be used for design of the shear reinforcement. In addition, where axial forces of tension or compression exist in the concrete frame, consideration must be given to the combined effects when designing for shear.

When torsional moments exist (twisting moments at right angles to the beam), their effects must be combined with beam shear.

Problem 7.4.A. A concrete beam similar to that shown in Figure 7.9 sustains a uniform live load of 1.5 klf and a uniform dead load of 1 klf on a span of 24 ft [7.32 m]. Determine the layout for a set of No. 3 U-stirrups using the stress method with f_y = 40 ksi [276 MPa] and f'_c = 3000 psi [20.7 MPa]. The beam section dimensions are b = 12 in. [305 mm] and d = 26 in. [660 mm].

Problem 7.4.B. Same as problem 7.4.A, except a span is 20 ft [6.1 m], b = 10 in. [254 mm], d = 23 in. [584 mm].

Problem 7.4.C. Determine the layout for a set of No. 3 U-stirrups for a beam with the same data as Problem 7.4.A, except the uniform live load is 0.75 klf [10.9 kN/m] and the uniform dead load is 0.5 klf [7.3 kN/m].

Problem 7.4.D. Determine the layout for a set of No. 3 U-stirrups for a beam with the same data as Problem 7.4.B, except the uniform live load is 1.875 klf [27.4 kN/m] and the uniform dead load is 1.25 klf [18.2 kN/m].

8

ANCHORAGE AND DEVELOPMENT OF REINFORCEMENT

The nature of reinforced concrete depends very primarily on the interactive relationship between the steel reinforcing bars and the concrete mass within which they are encased. Loads are basically applied to the concrete structure, that is, to the concrete mass. Stress developed in the steel must be accomplished through engagement between the steel and concrete, which occurs at their direct interface—the surface of the steel bars. In early uses of the stress method, the interfacing of the steel and concrete was visualized in terms of a stress at the bar surface, called *bond stress.* Bond stresses are developed on the surfaces of reinforcing bars whenever some structural action requires the steel and concrete to interact.

In times past, working stress procedures included the establishment of allowable stresses for bond and the computation of bond stresses for various situations. At present, however, the codes deal with this problem as one of development length. This chapter presents a discussion of bond stress situations and the current practices in establishing required lengths for the development of reinforcement.

8.1 DEVELOPMENT OF STRESS IN TENSION REINFORCEMENT

The ACI Code defines *development length* as the length of embedment required to develop the design strength of the reinforcement at a critical section. For beams, critical sections occur at points of maximum stress and at points within the span where some of the reinforcement terminates or is bent up or down. For a uniformly loaded, simple span beam, one critical section is at midspan, where the bending moment is at maximum. The tensile reinforcement required for flexure at this point must extend on both sides a sufficient distance to develop the stress in the bars; however, except for very short spans with large bars, the bar lengths will ordinarily be more than sufficient.

In the simple beam, the bottom reinforcement required for the maximum moment at midspan is not entirely required as the moment decreases toward the end of the span. It is thus sometimes the practice to make only part of the midspan reinforcement continuous for the whole beam length. In this case, it may be necessary to ensure that the bars that are of partial length are extended sufficiently from the midspan point, and that the bars remaining beyond the cutoff point can develop the stress required at that point.

When beams are continuous through the supports, top reinforcement is required for the negative moments at the supports. These top bars must be investigated for the development lengths, in terms of the distance they extend from the supports.

For tension reinforcement consisting of bars of No. 11 size and smaller, the code specifies a minimum length for development (L_d), as follows:

For No. 6 bars and smaller:

$$L_d = \frac{f_y d_b}{25\sqrt{f'_c}}$$

(but not less than 12 in.)

For No. 7 bars and larger:

$$L_d = \frac{f_y d_b}{20\sqrt{f'_c}}$$

In these formulas, d_b is the bar diameter.

Modification factors for L_d are given for various situations, as follows:

- For top bars in horizontal members with at least 12 in. of concrete below the bars: increase by 1.3.
- For flexural reinforcement that is provided in excess of that required by computations: decrease by a ratio of (required A_s/provided A_s).

Additional modification factors are given for lightweight concrete, for bars coated with epoxy, for bars encased in spirals, and for bars with f_y in excess of 60 ksi. The maximum value to be used for $\sqrt{f'_c}$ is 100 psi. Table 8.1 gives values for minimum development lengths for tensile reinforcement, based on the requirements of the ACI Code. The values listed under "other bars" are the unmodified length requirements; those listed under "top bars" are increased by the modification factor for this situation. Values are given for two concrete strengths and for the two most commonly used grades of tensile reinforcement.

The ACI Code makes no provision for a reduction factor for development lengths. As presented, the formulas for development length relate only to bar size, concrete strength, and steel yield strength. They are thus equally applicable for the stress method or the strength method with no further adjustment, except for the conditions previously described.

TABLE 8.1   Minimum Development Length for Tensile Reinforcement (in.)[a]

	$f_y = 40$ ksi [276 MPa]				$f_y = 60$ ksi [414 MPa]			
	$f'_c = 3$ ksi [20.7 MPa]		$f'_c = 4$ ksi [27.6 MPa]		$f'_c = 3$ ksi [20.7 MPa]		$f'_c = 4$ ksi [27.6 MPa]	
Bar Size	Top Bars[b]	Other Bars	Top Bars[b]	Other Bars	Top Bars[b]	Other Bars	Top Bars[b]	Other Bars
3	15	12	13	12	22	17	19	15
4	19	15	17	13	29	22	25	19
5	24	19	21	16	36	28	31	24
6	29	22	25	19	43	33	37	29
7	42	32	36	28	63	48	54	42
8	48	37	42	32	72	55	62	48
9	54	42	47	36	81	62	70	54
10	61	47	53	41	91	70	79	61
11	67	52	58	45	101	78	87	67

[a] Lengths are based on requirements of the ACI Code (Reference 1).

[b] Horizontal bars with more than 12 in. of concrete cast below them in the member.

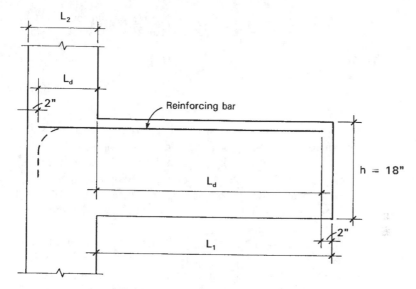

Figure 8.1   Reference for Example 1.

Example 1.   The negative moment in the short cantilever shown in Figure 8.1 is resisted by the steel bars in the top of the beam. Determine whether the development of the reinforcement is adequate without hooked ends on the No. 6 bars if L_1 = 48 in. [1220 mm] and L_2 = 36 in. [914 mm]. Use f'_c = 3 ksi [20.7 MPa] = and f_y = 60 ksi [414 MPa].

Solution: At the face of the support, anchorage for development must be achieved on both sides: within the support and in the top of the beam. In the top of the beam, the condition is one of "top bars," as previously defined. Thus, from Table 8.1, a length of 43 in. is required for L_d, which is adequately provided, if cover is minimum on the outside end of the bars.

   Within the support, the condition is one of "other bars" in the table reference. For this, the required length for L_d is 33 in., which is also adequately provided.

   Hooked ends are thus not required on either end of the bars, although most designers would probably hook the bars in the support, just for the security of the additional anchorage.

Problem 8.1.A.   A short cantilever is developed as shown in Figure 8.1. Determine whether adequate development is achieved without hooked ends on the bars if L_1 is 36 in. [914 mm], L_2 is 24 in. [610 mm], overall beam height is 16 in. [406 mm], bar size is No. 4, $f'_c = 4$ ksi [27.6 MPa], and $f_y = 40$ ksi [276 MPa].

Problem 8.1.B. Same as Problem 8.1.A, except $L_1 = 40$ in. [1020 mm], $L_2 = 30$ in. [762 mm], and bar size is No. 5.

8.2 HOOKS

When details of the construction restrict the ability to extend bars sufficiently to produce required development lengths, development can sometimes be assisted by use of a hooked end on the bar. So-called *standard hooks* may be evaluated in terms of a required development length, L_{dh}. Bar ends may be bent at 90°, 135°, or 180° to produce a hook. The 135° bend is used only for ties and stirrups, which normally consist of relatively small-diameter bars.

Table 8.2 gives values for development length with standard hooks, using the same variables for f'_c and f_y that are used in Table 8.1. The table values given are in terms of the required development length, as shown in Figure 8.2. Note that the table values are for 180° hooks, and that values may be reduced by 30% for 90° hooks. The following example illustrates the use of the data from Table 8.2 for a simple situation.

Example 2. For the bars in Figure 8.1, determine the length L_{dh} required for development of the bars with a 90° hooked end in the support. Use the same data as in Example 1.

TABLE 8.2 Required Development Length L_{dh} for Hooked Bars (in.)[a]

Bar Size	$f_y = 40$ ksi [276 MPa]		$f_y = 60$ ksi [414 MPa]	
	$f'_c = 3$ ksi [20.7 MPa]	$f'_c = 4$ ksi [27.6 MPa]	$f'_c = 3$ ksi [20.7 MPa]	$f'_c = 4$ ksi [27.6 MPa]
3	6	6	9	8
4	8	7	11	10
5	10	8	14	12
6	11	10	17	15
7	13	12	20	17
8	15	13	22	19
9	17	15	25	22
10	19	16	28	24
11	21	18	31	27

[a] See Figure 8.2. Table values are for a 180° hook; values may be reduced by 30% for a 90° hook.

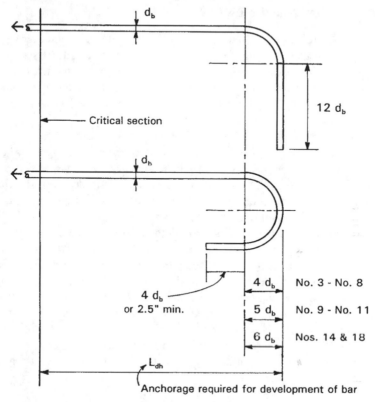

Figure 8.2   Detail requirements for standard hooks for use of values in Table 8.2.

Solution: From Table 8.2, the required length for the data given is 17 in. [432 mm] (No. 6 bar $f'_c = 3$ ksi, $f_y = 60$ ksi). This may be reduced for the 90° hook to

$$L = 0.70(17) = 11.9 \text{ in. [302 mm]}$$

Problem 8.2.A.   Find the development length required for the bars in Problem 8.1.A if the bar ends in the support are provided with 90° hooks.

Problem 8.2.B.   Find the development length required for the bars in Problem 8.1.B if the bar ends in the support are provided with 90° hooks.

8.3 BAR DEVELOPMENT IN CONTINUOUS BEAMS

Development length is the length of embedded reinforcement required to develop the design strength of the reinforcement at a critical section. Critical sections occur at points of maximum stress and at points within the span at which adjacent reinforcement terminates or is bent up into the top of the beam. For a uniformly loaded simple beam, one critical section is at midspan, where the bending moment is maximum. This is a point of maximum tensile stress in the reinforcement (peak bar stress), and some length of bar is required over which the stress can be developed. Other critical sections occur between midspan and the reactions, at points where some bars are cut off because they are no longer needed to resist the bending moment; such terminations create peak stress in the remaining bars that extend the full length of the beam.

When beams are continuous through their supports, the negative moments at the supports will require that bars be placed in the top of the beams. Within the span, bars will be required in the bottom of the beam for positive moments. Although the positive moment will go to zero at some distance from the supports, the codes require that some of the positive moment reinforcement be extended for the full length of the span and a short distance into the support.

Figure 8.3 shows a possible layout for reinforcement in a beam with continuous spans and a cantilevered end at the first support. Referring to the notation in the illustration:

1. Bars a and b are provided for the maximum moment of positive sign that occurs somewhere near the beam midspan. If all these bars are made full length (as shown for bars a), the length L_1 must be sufficient for development (this situation is seldom critical). If bars b are partial length, as shown in the illustration, then length L_2 must be sufficient to develop bars b, and length L_3 must be sufficient to develop bars a. As was discussed for the simple beam, the partial-length bars must actually extend beyond the theoretical cutoff point (B in the illustration), and the true length must include the dashed portions indicated for bars b.

2. For the bars at the cantilevered end, the distances L_4 and L_5 must be sufficient for development of bars c. L_4 is required to extend beyond the actual cutoff point of the negative moment by the extra length described for the partial-length bottom bars. If L_5 is not ade-

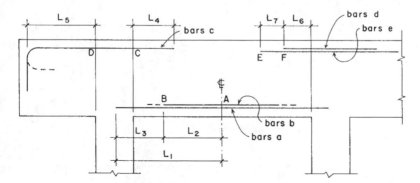

Figure 8.3   Development lengths in continuous beams.

quate, the bar ends may be bent into the 90° hook, as shown, or the 180° hook shown by the dashed line.

3. If the combination of bars shown in the illustration is used at the interior support, L_6 must be adequate for the development of bars d, and L_7 must be adequate for the development of bars e.

For a single loading condition on a continuous beam, it is possible to determine specific values of moment and their location along the span, including the locations of points of zero moment. In practice, however, most continuous beams are designed for more than a single loading condition, which further complicates the problems of determining the development lengths required.

8.4   SPLICES IN REINFORCEMENT

In various situations in reinforced concrete structures, it becomes necessary to transfer stress between steel bars in the same direction. Continuity of force in the bars is achieved by splicing, which may be effected by welding, by mechanical means, or by the lapped splice. Figure 8.4 illustrates the concept of the lapped splice, which consists essentially of the development of both bars within the concrete. Because a lapped splice is usually made with the two bars in contact, the lapped length must usually be somewhat greater than the simple development length required in Table 8.1.

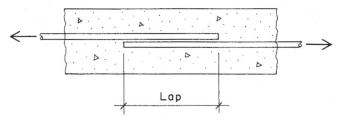

Figure 8.4 The lapped splice for steel reinforcing bars.

For a simple tension lapped splice, the full development of the bars usually requires a lapped length of 1.3 times that required for simple development of the bars. Lapped splices are generally limited to bars of No. 11 size or smaller.

For pure tension members, lapped splicing is not permitted, and splicing must be achieved by welding the bars or by some other mechanical connection. End-to-end butt welding of bars is usually limited to compression splicing of large diameter bars with high f_y, for which lapping is not feasible.

When members have several reinforcement bars that must be spliced, the splicing must be staggered. Splicing is generally not desirable and is to be avoided when possible, but because bars are obtainable only in limited lengths, some situations unavoidably involve splicing. Horizontal reinforcement in walls is one such case. For members with computed stress, splicing should not be located at points of maximum stress—for example, at points of maximum bending.

Splicing of compression reinforcement for columns is discussed in the next section.

8.5 DEVELOPMENT OF COMPRESSIVE REINFORCEMENT

Development length in compression is a factor in column design and in the design of beams reinforced for compression.

The absence of flexural tension cracks in the portions of beams where compression reinforcement is employed, along with the beneficial effect of the end bearing of the bars on the concrete, permit shorter developmental lengths in compression than in tension. The ACI Code prescribes that l_d for bars in compression shall be computed by the formula

$$L_d = \frac{0.02 f_y d_b}{\sqrt{f'_c}}$$

TABLE 8.3   Minimum Development Length for Compressive Reinforcement (in.)

Bar Size	$f_y = 40$ ksi [276 MPa]		$f_y = 60$ ksi [414 MPa]		
	$f'_c = 3$ ksi [20.7 MPa]	$f'_c = 4$ ksi [27.6 MPa]	$f'_c = 3$ ksi [20.7 MPa]	$f'_c = 4$ ksi [27.6 MPa]	$f'_c = 5$ ksi [34.5 MPa]
3	8	8	8	8	7
4	8	8	11	10	9
5	10	8	14	12	11
6	11	10	17	15	13
7	13	12	20	17	15
8	15	13	22	19	17
9	17	15	25	22	20
10	19	17	28	25	22
11	21	18	31	27	24
14			38	33	29
18			50	43	39

but shall not be less than $0.0003 f_y d_b$ or 8 in., whichever is greater. Table 8.3 lists compression bar development lengths for a few combinations of specification data.

In reinforced columns, both the concrete and the steel bars share the compression force. Ordinary construction practices require the consideration of various situations for development of the stress in the reinforcing bars. Figure 8.5 shows a multistory concrete column with its base supported on a concrete footing. With reference to the illustration, note the following:

1. The concrete construction is ordinarily produced in multiple, separate pours, with construction joints between the separate pours occurring as shown in the illustration.

2. In the lower column, the load from the concrete is transferred to the footing in direct compressive bearing at the joint between the column and footing. The load from the reinforcing must be developed by extension of the reinforcing into the footing: distance L_1 in the illustration. Although it may be possible to place the column bars in position during casting of the footing to achieve this, the common practice is to use dowels, as shown in the illustration. These dowels must be developed on both sides of the joint: L_1 in the foot-

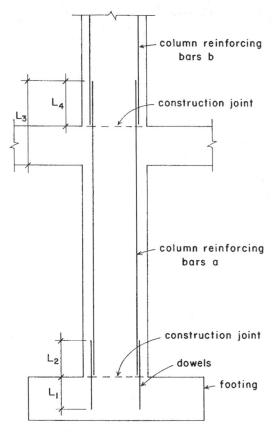

Figure 8.5 Development lengths in multistory columns.

ing and L_2 in the column. If the f'_c value for both the footing and the column are the same, these two required lengths will be the same.

3. The lower column will ordinarily be cast together with the supported concrete framing above it, with a construction joint occurring at the top level of the framing (bottom of the upper column), as shown in the illustration. The distance L_3 is that required to develop the reinforcing in the lower column—bars a in the illustration. As for the condition at the top of the footing, the distance L_4 is required to develop the reinforcing in bars b in the upper column. L_4 is more likely to be the critical consideration for the determination of the extension required for bars a.

8.6   DEVELOPED ANCHORAGE FOR FRAME CONTINUITY

In concrete rigid frame structures, the engagement of the frame members at joints between columns and beams requires special attention in the detailing of reinforcement. A particular concern is the potential for the beams to pull loose from the columns, an action typically resisted by the extended reinforcing bars from the beam ends. In addition to the usual concerns for bar development, some special detailing to enhance the anchoring of the bars may be indicated. This is a matter of particular note in resistance to seismic effects.

9

FLAT-SPANNING CONCRETE SYSTEMS

There are many different systems than can be used to achieve flat spans. These are used most often for floor structures, which typically require a dead flat form. However, in buildings with an all-concrete structure, they may also be used for roofs. Sitecast systems generally consist of one of the following basic types:

- One-way solid slab and beam
- Two-way solid slab and beam
- One-way joist construction
- Two-way flat slab or flat plate without beams
- Two-way joist construction, called *waffle construction*

Each system has its own distinct advantages and limits and some range of logical use, depending on required spans, general layout of supports, magnitude of loads, required fire ratings, and cost limits for design and construction.

The floor plan of a building and its intended usage determine loading conditions and the layout of supports. Also of concern are requirements for openings for stairs, elevators, large ducts, skylights, and so on, because these result in discontinuities in the otherwise commonly continuous systems. Whenever possible, columns and bearing walls should be aligned in rows and spaced at regular intervals in order to simplify design and construction and lower costs. However, the fluid concrete can be molded in forms not possible for wood or steel, and many very innovative, sculptural systems have been developed as takeoffs on these basic ones.

9.1 SLAB AND BEAM SYSTEMS

The most widely used and most adaptable cast-in-place, concrete floor system utilizes one-way solid slabs supported by one-way spanning beams. This system may be used for single spans, but it occurs more frequently with multiple-span slabs and beams in a system such as that shown in Figure 9.1.

In the example shown, the continuous slabs are supported by a series of beams that are spaced at 10 ft center to center. The beams, in turn, are supported by a girder-and-column system with columns at 30-ft centers, every third beam being supported directly by the columns, and the remaining beams being supported by the girders.

Because of the regularity and symmetry of the system shown in Figure 9.1, there are relatively few different elements in the basic system, each being repeated several times. Although special members must be designed for conditions that occur at the outside edge of the system and at the location of any openings for stairs, elevators, and so on, the general interior portions of the structure may be determined by designing only six basic elements: S1, S2, B1, B2, G1, and G2, as shown in the framing plan.

In computations for reinforced concrete, the span length of freely supported beams (simple beams) is generally taken as the distance between centers of supports or bearing areas; it should not exceed the clear span plus the depth of beam or slab. The span length for continuous or restrained beams is taken as the clear distance between faces of supports.

In continuous beams, negative bending moments are developed at the supports, and positive moments at or near midspan. This may be readily observed from the exaggerated deformation curve of Figure 9.2a. The exact values of the bending moments depend on several factors, but in the case of approximately equal spans supporting uniform loads, when the live

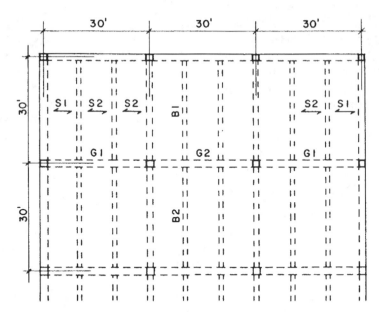

Figure 9.1 Framing layout for a typical slab and beam system.

load does not exceed three times the dead load, the bending moment values given in Figure 9.2 may be used for design.

The values given in Figure 9.2 are in general agreement with those given in Chapter 8 of the ACI Code. These values have been adjusted to account for partial live loading of multiple-span beams. Note that these values apply only to uniformly loaded beams. The ACI Code also gives some factors for end-support conditions other than the simple supports shown in Figure 9.2.

Design moments for continuous-span slabs are given in Figure 9.3. With large beams and short slab spans, the torsional stiffness of the beam tends to minimize the continuity effect in adjacent slab spans. Thus, most slab spans in the slab and beam systems tend to function much like individual spans with fixed ends.

Design of a One-Way Continuous Slab

The general design procedure for a one-way solid slab was illustrated in Section 6.7. The example given there is for a simple span slab. The following example illustrates the procedure for the design of a one-way continuous solid slab:

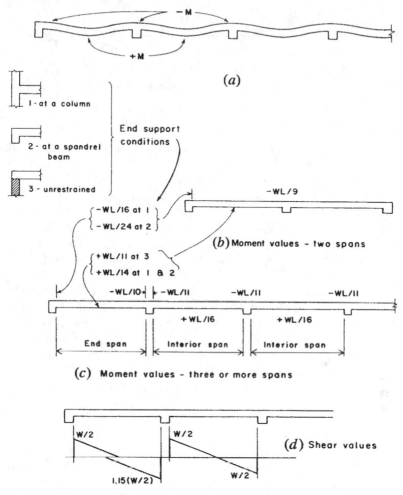

Figure 9.2   Approximate design factors for concrete beams.

Example 1.   A one-way solid slab is to be used for a framing system similar to that shown in Figure 9.1. Column spacing is 30 ft, with evenly spaced beams occurring at 10 ft center to center. Superimposed loads on the structure (floor live load plus other construction dead load) are a dead load of 38 psf [1.82 kPa] and a live load of 100 psf [4.79 kPa]. Use $f'_c =$ 3 ksi [20.7 MPa] and $f_y = 40$ ksi [276 MPa]. Determine the thickness for the slab and select its reinforcement.

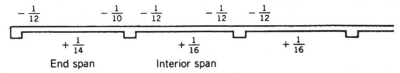

Figure 9.3 Approximate design factors for continuous slabs with spans of 10 ft or less.

Solution: To find the slab thickness, consider three factors: the minimum thickness for deflection, the minimum effective depth for the maximum moment, and the minimum effective depth for the maximum shear. For design purposes, the span of the slab is taken as the clear span, which is the dimension from face to face of the supporting beams. With the beams at 10-ft centers, this dimension is 10 ft, minus the width of one beam. Since the beams are not given, a dimension must be assumed for them. For this example, assume a beam width of 12 in., yielding a clear span of 9 ft.

Consider first the minimum thickness required for deflection. If the slabs in all spans have the same thickness (which is the most common practice), the critical slab is the end span because there is no continuity of the slab beyond the end beam. Although the beam will offer some restraint, it is best to consider this as a simple support; thus, the appropriate factor is $L/30$ from Table 6.8, and

$$\text{minimum } t = \frac{L}{30} = \frac{9 \times 12}{30} = 3.6 \text{ in. [91.4 mm]}$$

Assume here that fire-resistive requirements make it desirable to have a relatively heavy slab of 5-in. overall thickness, for which the dead weight of the slab is

$$w = \frac{5}{12} \times 150 = 62 \text{ psf [2.97 kPa]}$$

The total dead load is, thus, $62 + 38 = 100$ psf, and the factored total load is

$$U = 1.2(100) + 1.6(100) = 280 \text{ psf [13.4 kPa]}$$

Next, consider the maximum bending moment. Inspection of the moment values given in Figure 9.3 shows the maximum moment to be

$$M = \frac{1}{10}wL^2$$

With the clear span and the loading as determined, the maximum moment is, thus,

$$M = \frac{wL^2}{10} = \frac{280 \times (9)^2}{10} = 2268 \text{ ft-lb } [3.08 \text{ kN-m}]$$

and the required resisting moment for the slab is

$$M_R = \frac{2268}{0.9} = 2520 \text{ ft-lb } [3.42 \text{ kN-m}]$$

This moment value should now be compared to the balanced capacity of the design section, using the relationships discussed for rectangular beams in Section 6.3. For this computation, an effective depth for the design section must be assumed. This dimension will be the slab thickness minus the concrete cover and one-half the bar diameter. With the bars not yet determined, assume an approximate effective depth to be the slab thickness minus 1.0 in.; this will be exactly true with the usual minimum cover of 3/4 in. and a No. 4 bar. Then, using the balanced moment R factor from Table 6.1, the maximum resisting moment for the 12-in.-wide design section is

$$M_R = Rbd^2 = (1.149)(12)(4)^2 = 221 \text{ kip-in. } [25 \text{ kN-m}]$$

or

$$M_R = 221 \times \frac{1000}{12} = 18,400 \text{ ft-lb } [25 \text{ kN-m}]$$

Because this value is in excess of the required resisting moment of 2520 ft-lb, the slab is adequate for concrete flexural stress.

It is not practical to use shear reinforcement in one-way slabs, and consequently the maximum unit shear stress must be kept within the limit for the concrete alone. The usual procedure is to check the shear stress with the effective depth determined for bending before proceeding to find

A_s. Except for very short span slabs with excessively heavy loadings, shear stress is seldom critical.

For interior spans, the maximum shear will be $wL/2$, but for the end span it is the usual practice to consider some unbalanced condition for the shear due to the discontinuous end. Use a maximum shear of $1.15(wL/2)$, or an increase of 15% over the simple beam shear value. Thus,

$$\text{maximum shear} = V_u = 1.15 \times \frac{wL}{2} = 1.15 \times \frac{280 \times 9}{2}$$

$$= 1449 \text{ lb}[6.45 \text{ kN}]$$

and

$$\text{required } V_r = \frac{1449}{0.75} = 1932 \text{ lb } [8.59 \text{ kN}]$$

For the slab section with $b = 12$ in. and $d = 4$ in.:

$$V_c = 2\sqrt{f'_c}(b \times d) = 2\sqrt{3000}(12 \times 4) = 5258 \text{ lb } [23.4 \text{ kN}]$$

This is considerably greater than the required shear resistance, so the assumed slab thickness is not critical for shear stress.

Having thus verified the choice for the slab thickness, we may now proceed with the design of the reinforcement. For a balanced section, Table 6.1 yields a value of 0.685 for the a/d factor. However, since all sections will be classified as under-reinforced (actual moment less than the balanced limit), use an approximate value of 0.4 for a/d. Once the reinforcement for a section is determined, the true value of a/d can be verified, using the procedures developed in Section 6.3.

For the slab in this example, the following is computed:

$$\frac{a}{d} = 0.4, \quad \text{and} \quad a = 0.4(d) = 0.4(4) = 1.6 \text{ in. [40.6 mm]}$$

For the computation of required reinforcement, use

$$d - \frac{a}{2} = 4 - \frac{1.6}{2} = 3.2 \text{ in. [81.3 mm]}$$

Referring to Figure 9.3, note that there are five critical locations for which a moment must be determined and the required steel area computed. Reinforcement required in the top of the slab must be computed for the negative moments at the end support, at the first interior beam, and at the typical interior beam. Reinforcement required in the bottom of the slab must be computed for the positive moments at midspan locations in the first span and in the typical interior spans. The design for these conditions is summarized in Figure 9.4.

For the data displayed in the figure, note the following:

Maximum spacing of reinforcement:

$$s = 3t = 3(5) = 15 \text{ in. } [381 \text{ mm}]$$

Maximum required bending moment:

$$M_R = (\text{Moment factor } C)\left(\frac{wL^2}{0.9}\right) = C\left(\frac{280 \times (9)^2}{0.9}\right) \times 12 = 302,400C$$

Note that the use of the factor 12 gives this value for the moment in in.-lb units.

Required area of reinforcement.

$$A_s = \frac{M}{f_y\left(d - \dfrac{a}{2}\right)} = \frac{302,400 \times C}{40,000 \times 3.2} = 2.36C$$

Using data from Table 6.5, Figure 9.4 shows required spacing for Nos. 3, 4, and 5 bars. A possible choice for the slab reinforcement, using straight bars, is shown at the bottom of Figure 9.4.

For required temperature reinforcement

$$A_s = 0.002bt = 0.002(12 \times 5) = 0.12 \text{ in.}^2/\text{ft of slab width } [254 \text{ mm}^2/\text{m}]$$

Using data from Table 6.5, possible choices are #3 at 11 in. or #4 at 18 in.

Problem 9.1.A. A solid one-way slab is to be used for a framing system similar to that shown in Figure 9.1. Column spacing is 36 ft [11 m], with regularly spaced beams occurring at 12 ft [3.66 m] center to center. Superimposed dead

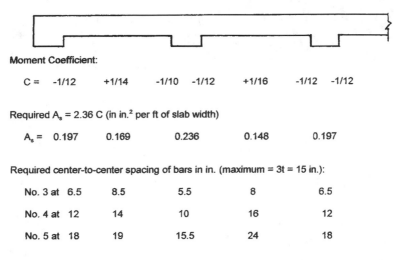

Moment Coefficient:

C =   -1/12   +1/14   -1/10   -1/12   +1/16   -1/12   -1/12

Required A$_s$ = 2.36 C (in in.^{2} per ft of slab width)

A$_s$ =   0.197   0.169   0.236   0.148   0.197

Required center-to-center spacing of bars in in. (maximum = 3t = 15 in.):

No. 3 at 6.5   8.5   5.5   8   6.5

No. 4 at 12   14   10   16   12

No. 5 at 18   19   15.5   24   18

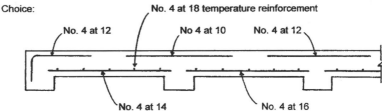

Choice:

Figure 9.4   Summary of design for the continuous slab.

load on the structure is 40 psf [1.92 kPa] and live load is 80 psf [3.83 kPa]. Use f'_c = 4 ksi [27.6 MPa] and f_y = 60 ksi [414 MPa]. Determine the thickness for the slab and select the size and spacing for the bars.

Problem 9.1.B.   Same as Problem 9.1.A, except that column spacing is 33 ft [10.1 m], beams are at 11 ft [3.35 m] centers, superimposed dead load is 50 psf [2.39 kPa], and live load is 75 psf [3.59 kPa].

9.2   GENERAL CONSIDERATIONS FOR BEAMS

The design of a single beam involves a large number of pieces of data, most of which are established for the system as a whole, rather than individually for each beam. System-wide decisions usually include those for the type of concrete and its design strength (f'_c), the type of reinforcing steel (f_y), the cover required for the necessary fire rating, and various generally used details for forming the concrete and placing the reinforcement. Most beams occur in conjunction with solid slabs that are cast monolithically with the beams. Slab thickness is established by the structural requirements of the spanning action between beams and by various concerns, such as those for fire rating, acoustic separation, type of reinforcement, and so on. Design of a single beam is usually limited to determination of the following:

1. Choice of shape and dimensions of the beam cross section
2. Selection of the type, size, and spacing of shear reinforcement
3. Selection of the flexural reinforcement to satisfy requirements based on the variation of moment along the several beam spans

The following are some factors that must be considered in effecting these decisions:

Beam Shape

Figure 9.5 shows the most common shapes used for beams in sitecast construction. The single, simple rectangular section is actually uncommon, but does occur in some situations. Design of the concrete section consists of selecting the two dimensions: the width b and the overall height or depth h.

As mentioned previously, beams occur most often in conjunction with monolithic slabs, resulting in the typical T-shape shown in Figure 9.5b or the L-shape shown in Figure 9.5c. The full T-shape occurs at the interior portions of the system, while the L shape occurs at the outside of the system or at the side of large openings. As shown in the illustration, there are four basic dimensions for the T and L that must be established in order to fully define the beam section:

- t = The slab thickness, which is ordinarily established on its own, rather than as a part of the single beam design
- h = The overall beam stem depth, corresponding to the same dimension for the rectangular section
- b_w = The beam stem width, which is critical for consideration of shear and for problems of fitting reinforcing into the section
- b_f = The so-called *effective width* of the flange, which is the portion of the slab assumed to work with the beam

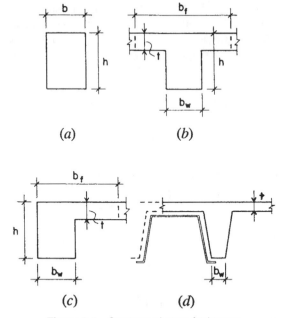

Figure 9.5 Common shapes for beams.

A special beam shape is that shown in Figure 9.5*d*. This occurs in concrete joist-and-waffle construction when "pans" of steel or reinforced plastic are used to form the concrete, the taper of the beam stem being required for easy removal of the forms. The smallest width dimension of the beam stem is ordinarily used for the beam design in this situation.

Beam Width

The width of a beam will affect its resistance to bending. Consideration of the flexure formulas given in Section 6.3 shows that the width dimension affects the bending resistance in a linear relationship (double the width and you double the resisting moment, etc.). On the other hand, the resisting moment is affected by the *square* of the effective beam depth. Thus, efficiency, in terms of beam weight or concrete volume, will be obtained by striving for deep, narrow beams, instead of shallow, wide ones— just as a 2 × 8 joist is more efficient that a 4 × 4 joist in wood.

Beam width also relates to various other factors, however, and these are often critical in establishing the minimum width for a given beam.

TABLE 9.1 Minimum Beam Widths[a]

Number of Bars	Bar Size								
	3	4	5	6	7	8	9	10	11
2	10	10	10	10	10	10	10	10	10
3	10	10	10	10	10	10	10	11	11
4	10	10	10	10	11	11	12	13	14
5	10	11	11	12	12	13	14	15	17
6	11	12	13	14	14	15	17	18	19
7	13	14	15	15	16	17	19	20	22
8	14	15	16	17	18	19	21	23	25
9	16	17	18	19	20	21	23	25	28
10	17	18	19	21	22	23	26	28	30

[a]Minimum width in inches for beams with 1.5-in. cover, No. 3 U-stirrups, clear spacing between bars of one bar diameter or minimum of 1 in. General minimum practical width for any beam with No. 3 U-stirrups is 10 in.

The formula for shear capacity indicates that the beam width is equally as effective as the depth in shear resistance. Placement of reinforcing bars is sometimes a problem in narrow beams. Table 9.1 gives minimum beam widths required for various bar combinations, based on considerations of bar spacing, minimum concrete cover of 1.5 in., placement of the bars in a single layer, and use of a No. 3 stirrup. Situations requiring additional concrete cover, use of larger stirrups, or the intersection of beams with columns may necessitate widths greater than those given in Table 9.1.

Beam Depth

Although selection of beam depth is partly a matter of satisfying structural requirements, it is typically constrained by other considerations in the building design. Figure 9.6 shows a section through a typical building floor/ceiling with a concrete slab and beam structure. In this situation, the critical depth from a general building design point of view is the overall thickness of the construction, shown as H in the illustration. In addition to the concrete structure, this includes allowances for the floor finish, the ceiling construction, and the passage of an insulated air duct. The net usable portion of H for the structure is shown as the dimension h, with the effective structural depth d being something less than h. Because the space defined by H is not highly usable for the building occupancy, there is a tendency to constrain it, which works to limit any extravagant use of d.

Most concrete beams tend to fall within a limited range in terms of the ratio of width to depth. The typical range is for a width-to-depth ratio

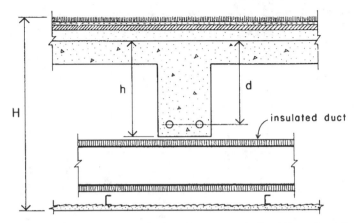

Figure 9.6 Concrete beam in typical multistory construction. Dimension *H* is critical for architectural planning; dimension *d* is critical for structural design.

between 1:1.5 and 1:2.5, with an average of 1:2. This is not a code requirement or a magic rule; it is merely the result of satisfying typical requirements for flexure, shear, bar spacing, economy of use of steel, and deflection.

Deflection Control

Deflection of spanning slabs and beams must be controlled for a variety of reasons. This topic is discussed in Section 6.8. Typically, the most critical decision factor relating to deflection is the overall vertical thickness or height of the spanning member. The ratio of this height dimension to the span length is the most direct indication of the degree of concern for deflection.

Design of Continuous Beams

Continuous beams are typically indeterminate and must be investigated for the bending moments and shears that are critical for the various loading conditions. When the beams are not involved in rigid-frame actions (as they are when they occur on column lines in multistory buildings), it may be possible to use approximate analysis methods, as described in the ACI Code and here in Section 9.1. An illustration of such a procedure is shown in the design of a concrete floor structure in Chapter 16.

In contrast to beams of wood and steel, those of concrete must be designed for the changing internal force conditions along their length. The single, maximum values for bending moment and shear may be critical in establishing the required beam size, but requirements for reinforcement must be investigated at all supports and midspan locations.

9.3 OTHER FLAT-SPANNING SYSTEMS

While the slab and beam system is the most used and most adaptable site-cast framing system, there are other common systems that may be used when circumstances permit. The following is a description of some of these systems:

One-Way Joist Construction

Figure 9.7 shows a partial framing plan and some details for a type of construction that utilizes a series of very closely spaced beams and a relatively thin solid slab. Because of its resemblance to ordinary wood joist construction, this is called *concrete joist construction.* This system is generally the lightest (in dead weight) of any type of flat-spanning, site-cast concrete construction and is structurally well suited to the light loads and medium spans of office buildings and commercial retail buildings. Although popular in times past, the lack of fire resistance of this construction now makes it somewhat less attractive.

Slabs as thin as 2 in. and joists as narrow as 4 in. are used with this construction. Because of the thinness of the parts and the small amount of cover provided for reinforcement (typically ¾ to 1 in. for joists versus 1.5 in. for ordinary beams), the construction has very low resistance to fire, especially when exposed from the underside. It is therefore necessary to provide some form of fire protection, as for steel construction, or to restrict its use to situations where high fire ratings are not required.

The relatively thin, short-span slabs are typically reinforced with welded wire mesh rather than ordinary deformed bars. Joists are often tapered at their ends, as shown in the framing plan in Figure 9.7. This is done to provide a larger cross section for increased resistance to shear and negative moment at the supports. Shear reinforcement in the form of single vertical bars may be provided, but is not frequently used.

Early joist construction was produced by using lightweight, hollow, clay tile blocks to form the voids between joists. These blocks were simply arranged in spaced rows on top of the forms, the joists being formed by the

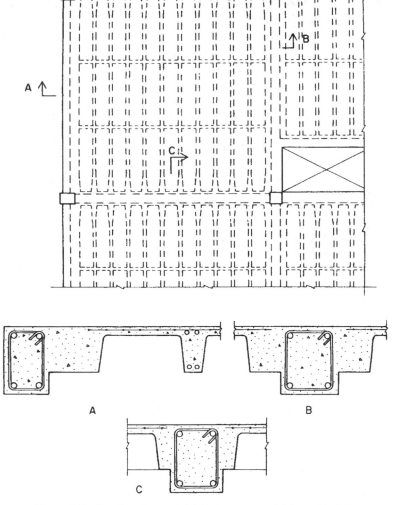

Figure 9.7 Framing plan and details for one-way joist construction.

spaces between the rows. The resulting construction provided a flat underside to which a plastered ceiling surface could be directly applied. Hollow, lightweight concrete blocks later replaced the clay tile blocks. Other
forming systems have utilized plastic-coated cardboard boxes, fiber-
glass-reinforced pans, and formed sheet-metal pans. The latter method

was very widely used, the metal pans being pried off after the pouring of the concrete and reused for several additional pours. The tapered joist cross section shown in Figure 9.7 is typical of this construction, because the removal of the metal pans requires it.

Wider joists can be formed by simply increasing the space between forms (individual rows of pans), with large beams being formed in a similar manner or by the usual method of extending a beam stem below the construction, as shown for the beams in Figure 9.7. Because of the usual narrow joist shape, cross-bridging is usually required, just as with wood joist construction. The framing plan in Figure 9.7 shows the use of two bridging strips in the typical bay of the framing.

Design of joist construction is essentially the same as for ordinary slab and beam construction. Some special regulations are given in the ACI Code for this construction, such as the reduced cover mentioned previously. Because joists are so commonly formed with standard-sized metal forms, there are tabulated designs for typical systems in various handbooks. The *CRSI Handbook* (Reference 5) has extensive tables offering complete designs for various spans, loadings, pan sizes, and so on. Whether for final design or simply for a quick preliminary design, the use of such tables is quite efficient.

One-way joist construction was highly popular in earlier times, but has become less utilized, due to its lack of fire resistance and the emergence of other systems. The popularity of lighter, less fire-resistive ceiling construction has been a contributing factor, as well as the development of various prestressed and precast systems. In the right situation, however, joist construction is still a highly efficient type of construction.

Waffle Construction

Waffle construction consists of two-way spanning joists that are formed in a manner similar to that for one-way spanning joists, using forming units of metal, plastic, or cardboard to produce the void spaces between the joists. The most widely used type of waffle construction is the waffle flat slab, in which solid portions around column supports are produced by omitting the void-making forms. An example of a portion of such a system is shown in Figure 9.8. This type of system is analogous to the solid flat slab. At points of discontinuity in the plan—such as at large openings or at edges of the building—it is usually necessary to form beams. These beams may be produced as projections below the waffle, as shown in Figure 9.8, or may be created within the waffle depth by omitting a row of the void-making forms, as shown in Figure 9.9.

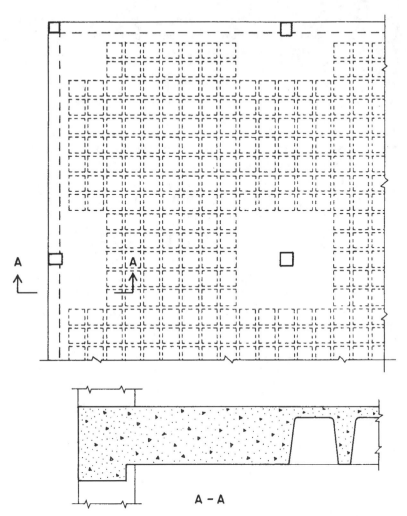

Figure 9.8 Framing plan and details for a waffle system (two-way spanning joists) with no interior column-line beams.

If beams are provided on all of the column lines, as shown in Figure 9.9, the construction is analogous to the two-way solid slab with edge supports. With this system, the solid portions around the column are not required because the waffle itself does not achieve the transfer of high shear or development of the high negative moments at the columns.

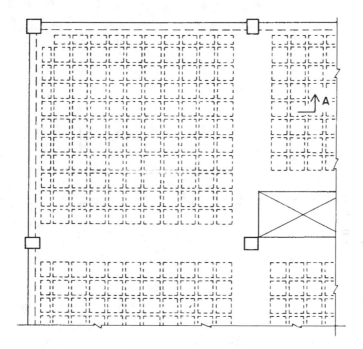

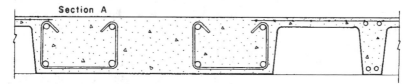

Figure 9.9 Framing plan and details for a waffle system with interior column-line beams developed within the waffle system.

As with the one-way joist construction, fire ratings are low for ordinary waffle construction. The system is best suited for situations involving relatively light loads, medium-to-long spans, approximately square column bays, and a reasonable number of multiple bays in each direction.

For the waffle construction shown in Figure 9.8, the edge of the structure represents a major discontinuity when the column supports occur immediately at the edge, as shown. Where planning permits, a more efficient use of the system is represented by the partial framing plan shown

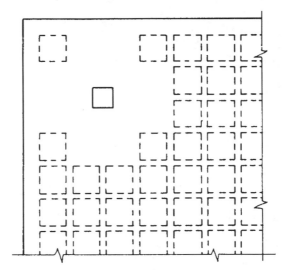

Figure 9.10 Framing plan detail for a waffle system with cantilevered edges and no column-line beams.

in Figure 9.10, in which the edge occurs some distance past the columns. This projected edge provides a greater shear periphery around the column and helps to generate a negative moment, preserving the continuous character of the spanning structure. With the use of the projected edge, it may be possible to eliminate the edge beams shown in Figure 9.8, thus preserving the waffle depth as a constant.

Another variation for the waffle is the blending of some one-way joist construction with the two-way waffle joists. This may be achieved by keeping the forming the same as for the rest of the waffle construction and merely using the ribs in one direction to create the spanning structure. One reason for doing this is the situation shown in Figure 9.9, where the large opening for a stair or elevator results in a portion of the waffle (the remainder of the bay containing the opening) being considerably out of square, that is, having one span considerably greater than the other. The joists in the short direction in this case will tend to carry most of the load due to their greater stiffness (less deflection than the longer spanning joists that intersect them). Thus, the short joists would be designed as one-way spanning members, and the longer joists would have only minimum reinforcing and serve as bridging elements.

The two-way spanning waffle systems are quite complex in structural behavior, and their investigation and design are beyond the scope of this book. Some aspects of this construction are discussed in the next section because there are many similarities between the two-way spanning waffle systems and the two-way spanning solid slab systems. As with the one-way joist system, there are some tabulated designs in various handbooks that may be useful for either final or preliminary design. The *CRSI Handbook* (Reference 5) mentioned previously has some such tables.

For all two-way construction, such as the waffle system, feasible use of the system depends on the logical development of the general building plans, regarding the arrangements of structural supports, locations of openings, length of spans, and so on. In the right situation, these systems may be able to realize their full potential, but if order, symmetry, and other factors are lacking—resulting in major adjustments away from the simple two-way functioning of the system—it may be very unreasonable to select such a structure. In some cases, the waffle has been chosen strictly for its underside appearance; it has also been pushed into use in situations not fitted to its nature. There may be some justification for such cases, but the resulting structures are likely to be quite awkward.

An illustration of the use of waffle construction for a roof structure is presented in the building design example in Section 16.8.

Two-Way Spanning Solid-Slab Construction

If reinforced in both directions, the solid concrete slab may span in two ways as well as one. The widest use of such a slab is in flat-slab or flat-plate construction. In flat-slab construction, beams are used only at points of discontinuity, with the typical system consisting only of the slab and the strengthening elements used at column supports. Typical details for a flat-slab system are shown in Figure 9.11.

Drop panels consisting of thickened portions square in plan are used to give additional resistance to the high shear and negative moment that develop at the column supports. Enlarged portions are also sometimes provided at the tops of the columns (called *column capitals*) to reduce the stresses in the slab further.

Two-way slab construction consists of multiple bays of two-way solid spanning slabs with edge supports consisting of bearing walls of concrete or masonry or of column-line beams formed in the usual manner. Typical details for such a system are shown in Figure 9.12.

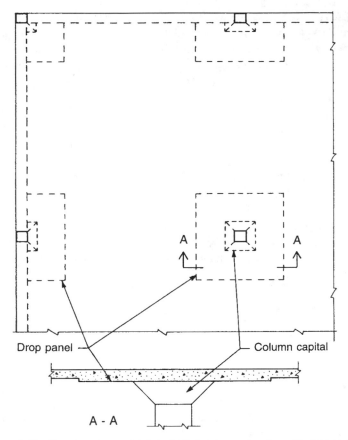

Drop panel

Column capital

A - A

Figure 9.11 Framing plan and details for flat-slab construction with drop panels (thickened slab) and column capitals (flared tops).

Two-way solid-slab construction is generally favored over waffle construction where higher fire rating is required for the unprotected structure or where spans are short and loadings high. As with all types of two-way spanning systems, solid-slab systems function most efficiently where the spans in each direction are approximately the same.

For investigation and design, the flat slab (Figure 9.11) is considered to consist of a series of one-way spanning solid-slab strips. Each of these strips spans through multiple bays in the manner of a continuous beam and is supported either by columns or by the strips that span in a direction perpendicular to it. The analogy for this is shown in Figure 9.13a.

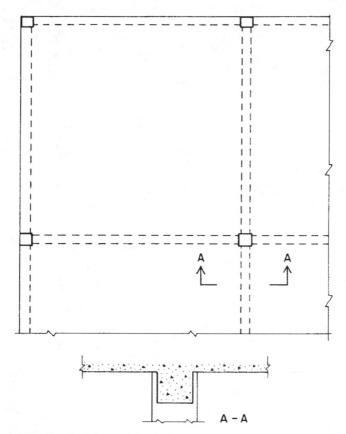

Figure 9.12 Framing plan and details for a two-way slab with edge supports (column-line beams or bearing walls).

As shown in Figure 9.13*b*, the slab strips are divided into two types: those passing over the columns, and those passing between columns, called *middle strips*. The complete structure consists of the intersecting series of these strips, as shown in Figure 9.13*c*. For the flexural action of the system, there is two-way reinforcing in the slab at each of the boxes defined by the intersections of the strips. In box 1 in Figure 9.13*c*, both sets of bars are in the bottom portion of the slab, due to the positive moment in both intersecting strips. In box 2, the middle-strip bars are in the top (for negative moment), while the column-strip bars are in the bottom (for positive moment). In box 3, the bars are in the top in both directions.

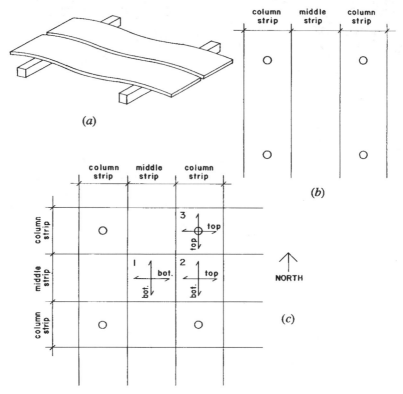

Figure 9.13 The two-way spanning flat slab visualized as a series of column strips and middle strips.

Composite Construction: Concrete with Structural Steel

Figure 9.14 shows a section detail of a type of construction generally referred to as *composite construction*. This consists of a sitecast concrete spanning slab supported by structural steel beams, the two being made to interact by the use of shear developers welded to the top of the beams and embedded in the cast slab. The concrete may be formed by plywood sheets placed against the underside of the beam flange, resulting in the detail shown in Figure 9.14.

A variation on the forming shown in Figure 9.14 consists of using light-gauge formed sheet steel decking to support the cast concrete. The shear developers are then site-welded through the thin deck to the tops of the beams.

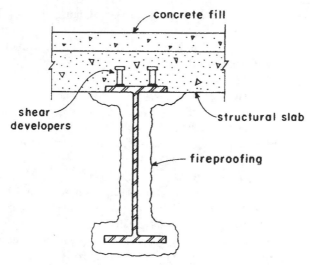

Figure 9.14 Composite construction with steel beams and a sitecast concrete slab.

Although it is common to refer to this form of construction as *composite construction,* in its true meaning the term covers any situation in which more than a single material is made to develop a singular structural response. By the general definition, therefore, even ordinary reinforced concrete is "composite," with the concrete interacting with the steel reinforcement. Other examples include laminated glazing (glass plus plastic), flitched beams (wood plus steel), and concrete fill on top of steel deck when the deck is bonded to the concrete.

The AISC manual (*Manual of Steel Construction,* published by the American Institute of Steel Construction) contains data and design examples for the type of construction shown in Figure 9.14.

Precast Construction

Precast construction components may be used to produce entire structural systems, but are more frequently used in combination with other structural components, such as the following:

- *Precast Decks.* These may be solid slabs, hollow-cored slabs, or various contoured, ribbed slabs. These decks can be supported by sitecast concrete, masonry walls, or steel frames.

- *Tilt-Up Walls.* These are routinely used with horizontal systems for roofs and floors consisting of wood or steel framing. However, they can also be combined with sitecast concrete or with other elements of precast concrete.

Flat-spanning systems can also be produced by using modular units of precast concrete to form sitecast systems, such as one-way joists or waffles. The exposed undersides of such construction are able to be achieved with finer detail and finish quality than with just about any other way of forming.

Some very imaginative structures have been developed with the use of custom-designed, complete systems of precast concrete. This generally demands a considerably greater design effort and considerable coordination of the design and production work. This use also requires real dedication on the part of everyone involved in the complete building design, to accomplish something far beyond the routine work of construction.

Use of Prestressing

There are various advantages to the use of prestressing in flat-spanning structures. A principal one is the reduction of creep deflections and cracking, the latter being significant when the underside of the construction is exposed to view. Prestressing is used generally with flat-spanning precast concrete components, but can also be used with some sitecast systems in the form of post-tensioning.

General considerations for the use of prestressing are discussed in Section 1.7.

9.4 DESIGN AIDS

The design of various elements of reinforced concrete can be aided—or, in many cases, totally achieved—by the use of various prepared materials. Handbooks present complete data for various elements, such as footings, columns, one-way slabs, joist construction, waffle systems, and two-way slab systems. For the design of a single footing or a one-way slab, the handbook merely represents a convenience or a shortcut to a final design. For columns subjected to bending, for waffle construction, and for two-way slab systems, "longhand" design (without aid, except from a pocket calculator) is really not feasible. In the latter cases, handbook data may be used to establish a reasonable preliminary design, which may then be custom-fit to the specific conditions by some investigation and computations. Even the largest of handbooks cannot present all possible combinations of values of f'_c, grade of reinforcing bars, value of super-imposed loads, and so on. Thus, only coincidentally will handbook data be exactly correct for any specific design job.

In the age of the computer, there is a considerable array of software available for the routine tasks of structural design. For many of the complex and laborious problems of design of reinforced concrete structures, these software options are a real boon for anyone able to utilize them.

10

CONCRETE COLUMNS

In view of the ability of concrete to resist compressive stress and its weakness in tension, it would seem to be apparent that its most logical use is for structural members whose primary task is the resistance of compression. This observation ignores the use of reinforcement to a degree, but is nevertheless noteworthy. And, indeed, major use is made of concrete for columns, piers, pedestals, posts, and bearing walls—all basically compression members. This chapter presents discussions of the use of reinforced concrete for such structural purposes, with emphasis on the development of columns for building structures. Concrete columns often exist in combination with concrete beam systems, forming rigid frames with vertical planar bents. This subject is addressed in Chapter 11 and as part of the discussion of the concrete example structure in Chapter 16.

10.1 EFFECTS OF COMPRESSION FORCE

When concrete is subjected to a direct compressive force, the most obvious stress response in the material is one of compressive stress, as shown in Figure 10.1a. This response may be the essential concern, as

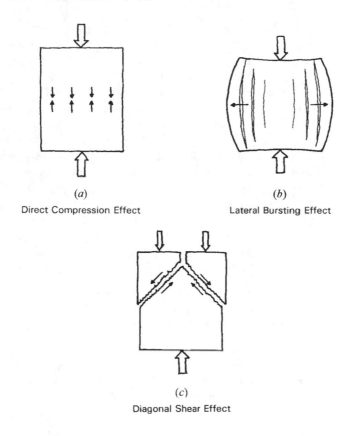

Figure 10.1   Fundamental failure modes of the tension-weak concrete.

it would be in a wall composed of flat, precast concrete bricks stacked on top of each other. Direct compressive stress in the individual bricks and in the mortar joints between bricks would be a primary situation for investigation.

However, if the concrete member being compressed has some dimension in the direction of the compressive force—as in the case of a column or pier—there are other internal stress conditions that may well be the source of structural failure under the compressive force. Direct compressive force produces a three-dimensional deformation that includes a pushing out of the material at right angles to the force, actually producing tension stress in that direction, as shown in Figure 10.1*b*. In a tension-weak material, this tension action may produce a lateral bursting effect.

Because concrete as a material is also weak in shear, another possibility for failure is along the internal planes where maximum shear stress is developed. This occurs at a 45° angle with respect to the direction of the applied force, as shown in Figure 10.1c.

In concrete compression members, other than flat bricks, it is generally necessary to provide for all three stress responses shown in Figure 10.1. In fact, additional conditions can occur if the structural member is also subjected to bending or torsional twisting. Each case must be investigated individually for all the individual actions and the combinations in which they can occur. Design for the concrete member and its reinforcement will typically respond to several considerations of behavior, and the same member and reinforcement must function for all responses. The following discussions focus on the primary function of resistance to compression, but other concerns will also be mentioned.

The basic consideration for combined compression and bending is discussed here because current codes require that all columns be designed for this condition.

10.2 REINFORCEMENT FOR COLUMNS

Column reinforcement takes various forms and serves various purposes, the essential consideration being to enhance the structural performance of the column. Considering the three basic forms of column stress failure shown in Figure 10.1, it is possible to visualize basic forms of reinforcement for each condition. This is done in the illustrations in Figures 10.2a, b, and c.

To assist the basic compression function, steel bars are added, with their linear orientation in the direction of the compression force. This is the fundamental purpose of the vertical reinforcing bars in a column. Although the steel bars displace some concrete, their superior strength and stiffness make them a significant improvement.

To assist in resistance to lateral bursting (Figure 10.2b), a critical function is to hold the concrete from moving out laterally, which may be achieved by so-called *containment* of the concrete mass, similar to the action of a piston chamber containing air or hydraulic fluid. If compression resistance can be obtained from air that is contained, surely it can be more significantly obtained from contained concrete. This is a basic reason for the extra strength of the traditional spiral column, and one reason that very closely spaced ties in tied columns are now favored. In retrofitting columns for improved seismic resistance, a technique sometimes

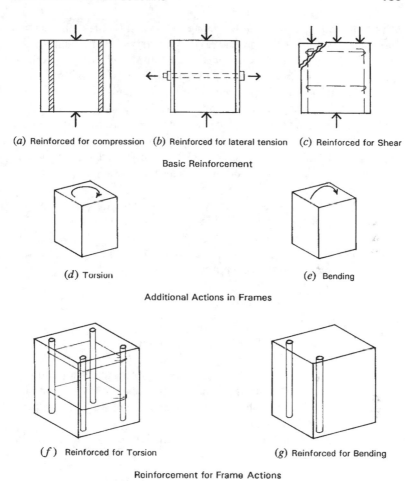

(a) Reinforced for compression (b) Reinforced for lateral tension (c) Reinforced for Shear

Basic Reinforcement

(d) Torsion (e) Bending

Additional Actions in Frames

(f) Reinforced for Torsion (g) Reinforced for Bending

Reinforcement for Frame Actions

Figure 10.2 Forms and functions of column reinforcement.

used is to actually provide a confining exterior jacket of steel or fiber strand, essentially functioning as illustrated in Figure 10.2b.

Natural shear resistance is obtained from the combination of the vertical bars and the lateral ties or spiral, as shown in Figure 10.2c. If this is a critical concern, improvements can be obtained by using closer-spaced ties and a larger number of vertical bars that spread out around the column perimeter.

When used as parts of concrete frameworks, columns are also typically subjected to torsion and bending, as shown in Figures 10.2d and e. Torsional twisting tends to produce a combination of longitudinal tension and lateral shear; thus, the combination of full-perimeter ties or spirals and the perimeter vertical bars provide for this in most cases.

Bending, if viewed independently, requires tension reinforcement, just as in an ordinary beam. In the column, the ordinary section is actually a doubly reinforced one, with both tension and compression reinforcement for beam action. This function, combined with the basic axial compression, is discussed more fully in later sections of this chapter. An added complexity in many situations is the existence of bending in more than a single direction.

All of these actions can occur in various combinations due to different conditions of loading. Column design is, thus, a quite complex process when all possible structural functions are considered. The need to make multiple usage of the simplest combination of reinforcing elements becomes a fundamental design principle.

10.3 TYPES OF COLUMNS

Concrete columns occur most often as the vertical support elements in a structure generally built of cast-in-place concrete (commonly called *site cast*). This situation is discussed in this chapter. Very short columns, called *pedestals,* are sometimes used in the support system for columns or other structures. The ordinary pedestal is discussed as a foundation transitional device in Chapter 13. Walls that serve as vertical compression supports are called *bearing walls.*

The sitecast concrete column usually falls into one of the following categories:

- Square columns with tied reinforcement
- Oblong columns with tied reinforcement
- Round columns with tied reinforcement
- Round columns with spiral-bound reinforcement
- Square columns with spiral-bound reinforcement
- Columns of other geometries (L-shaped, T-shaped, octagonal, etc.) with either tied or spiral-bound reinforcement

Obviously, the choice of column cross-sectional shape is an architectural, as well as a structural, decision. However, forming methods and costs, arrangement and installation of reinforcement, and relations of the column form and dimensions to other parts of the structural system must also be dealt with.

In tied columns, the longitudinal reinforcement is held in place by loop ties made of small-diameter reinforcement bars, commonly No. 3 or No. 4 bars. Such a column is represented by the square section shown in Figure 10.3a. This type of reinforcement can quite readily accommodate other geometries, as well as the square.

Spiral columns are those in which the longitudinal reinforcing is placed in a circle, with the whole group of bars enclosed by a continuous cylindrical spiral made from steel rod or large-diameter steel wire. Although this reinforcing system obviously works best with a round column section, it can be used with other geometries also. A round column of this type is shown in Figure 10.3b.

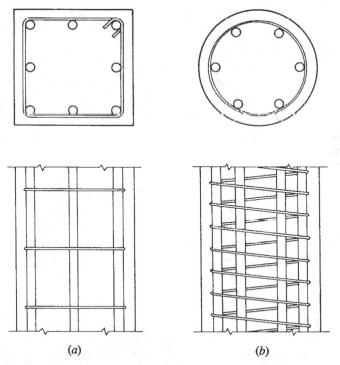

(a)   (b)

Figure 10.3 Primary forms of column reinforcement: (a) rectangular layout of vertical bars with lateral ties, and (b) circular layout of vertical bars with continuous helix (spiral) wrap.

Experience has shown the spiral column to be slightly stronger than an equivalent tied column with the same amount of concrete and reinforcement. For this reason, code provisions have traditionally allowed slightly more load on spiral columns. Spiral reinforcement tends to be expensive, however, and the round bar pattern does not always mesh well with other construction details in buildings. Thus, tied columns are often favored where restrictions on the outer dimensions of the sections are not severe.

A recent development is the use of tied columns with very closely spaced ties. A basic purpose for this is the emulation of a spiral column, for achieving additional strength, although many forms of gain are actually obtained simultaneously, as discussed in regard to the illustrations in Figure 10.2.

10.4 GENERAL REQUIREMENTS FOR COLUMNS

Code provisions and practical construction considerations place a number of restrictions on column dimensions and choice of reinforcement.

Column Size. The current code does not contain limits for column dimensions. For practical reasons, the following limits are recommended. Rectangular tied columns should be limited to a minimum area of 100 in.^{2} and a minimum side dimension of 10 in. if square and 8 in. if oblong. Spiral columns should be limited to a minimum size of 12 in. if either round or square.

Reinforcement. Minimum bar size is No. 5. The minimum number of bars is four for tied columns, five for spiral columns. The minimum amount of area of steel is 1% of the gross column area. A maximum area of steel of 8% of the gross area is permitted, but bar spacing limitations makes this difficult to achieve; 4% is a more practical limit. The ACI Code stipulates that for a compression member with a larger cross section than required by considerations of loading, a reduced effective area not less than one-half the total area may be used to determine minimum reinforcement and design strength.

Ties. Ties should be at least No. 3 for bars that are No. 10 and smaller. No. 4 ties should be used for bars that are No. 11 and larger. Vertical spacing of ties should be not more than 16 times the vertical bar diameter, 48 times the tie diameter, or the least dimension of the column. Ties should be arranged so that every corner and alternate longitudinal bar is held by the corner of a tie with an included angle of not greater than 135°, and no bar should be farther than 6 in. clear from such a supported bar. Complete circular ties may be used for bars placed in a circular pattern.

Concrete Cover. A minimum of 1.5 in. cover is needed when the column surface is not exposed to weather and is not in contact with the ground. Cover of 2 in. should be used for formed surfaces exposed to the weather or in contact with the ground. Cover of 3 in. should be used if the concrete is cast directly against earth without constructed forming, such as occurs on the bottoms of footings.

Spacing of Bars. Clear distance between bars should not be less than 1.5 times the bar diameter, 1.33 times the maximum specified size for the coarse aggregate, or 1.5 in.

10.5 COMBINED COMPRESSION AND BENDING

Because of the nature of most concrete structures, design practices generally do not consider the possibility of a concrete column with axial compression alone. This is to say, the existence of some bending moment is always considered, together with the axial force. Figure 10.4 illustrates the nature of the so-called *interaction response* for a concrete column, with a range of combinations of axial load plus bending moment. In general, there are three basic ranges of this behavior, as follows (see the dashed lines in Figure 10.4):

1. *Large Axial Force, Minor Moment.* For this case, the moment has little effect, and the resistance to pure axial force is only negligibly reduced.
2. *Significant Values for Both Axial Force and Moment.* For this case, the analysis for design must include the full combined force effects, that is, the interaction of the axial force and the bending moment.
3. *Large Bending Moment, Minor Axial Force.* For this case, the column behaves essentially as a doubly reinforced (tension and compression reinforced) member, with its capacity for moment resistance affected only slightly by the axial force.

In Figure 10.4, the solid line on the graph represents the true response of the column—a form of behavior verified by many laboratory tests.

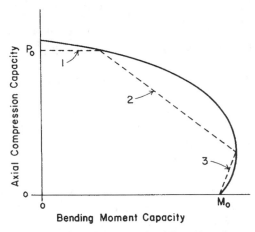

Figure 10.4   Interaction of axial compression (P) and bending moment (M) in a reinforced concrete column. Solid line indicates general form of response; dashed line indicates three separate zones of response: (1) dominant compression with minor bending, (2) significant compression plus bending interaction, and (3) dominant bending in the cracked section.

The dashed line represents the generalization of the three types of response just described.

The terminal points of the interaction response—pure axial compression or pure bending moment—may be reasonably easily determined (P_o and M_o in Figure 10.4). The interaction responses between these two limits require complex analyses beyond the scope of this book.

A special type of bending is generated when a relatively slender compression member develops some significant curvature due to the bending induced at its ends. In this case, the center portion of the member's length literally moves away (deflects sideways; called *delta* for the deflected dimension) from a straight line. In this center portion, therefore, a bending is induced, consisting of the product of the compression force (P) and the deflected dimension (delta). As this effect produces additional bending, and thus additional deflection, it may become a progressive failure condition—as it indeed is for very slender members. This is called the *P-delta effect,* and it is a critical consideration for relatively slender columns. Because concrete columns are not usually very slender, this effect is usually of less concern than it is for columns of wood and steel.

10.6 CONSIDERATIONS FOR COLUMN SHAPE

Usually, a number of possible combinations of reinforcing bars may be assembled to satisfy the steel area requirement for a given column. Aside from providing for the required cross-sectional area, the number of bars must also work reasonably in the layout of the column. Figure 10.5 shows a number of columns with various numbers of bars. When a square tied column is small, the preferred choice is usually the simple four-bar layout, with one bar in each corner and a single perimeter tie. As the column gets larger, the distance between the corner bars gets larger, and it is best to use more bars so that the reinforcement is spread out around the column periphery. For a symmetrical layout and the simplest of tie layouts, the best choice is for numbers that are multiples of four, as shown in Figure 10.5*a*. The number of additional ties required for these layouts depends on the size of the column and the considerations discussed in Section 10.4.

An unsymmetrical bar arrangement (Figure 10.5*b*) is not necessarily bad, even though the column and its construction details are otherwise not oriented differently on the two axes. In situations where moments may be greater on one axis, the unsymmetrical layout is actually preferred; in fact, the column shape will also be more effective if it is unsymmetrical, as shown for the oblong shapes in Figure 10.5*c*.

Figures 10.5*d* through *g* show a number of special column shapes developed as tied columns. Although spirals could be used in some cases for such shapes, the use of ties allows much greater flexibility and simplicity of construction. One reason for using ties may be the column dimensions, there being a practical lower limit of about 12 in. in width for a spiral-bound column.

Round columns are frequently formed as shown in Figure 10.5*h*, if built as tied columns. This allows for a minimum reinforcement with four bars. If a round pattern is used (as it must be for a spiral-bound column), the usual minimum number recommended is six bars, as shown in Figure 10.5*i*. Spacing of bars is much more critical in spiral-bound circular arrangements, making it very difficult to use high percentages of steel in the column section. For very-large-diameter columns, it is possible to use sets of concentric spirals, as shown in Figure 10.5*j*.

For cast-in-place columns, a concern that must be dealt with is the vertical splicing of the steel bars. Two places where this commonly occurs are at the top of the foundation and at floors where a multistory column continues upward. At these points, there are three ways to achieve the

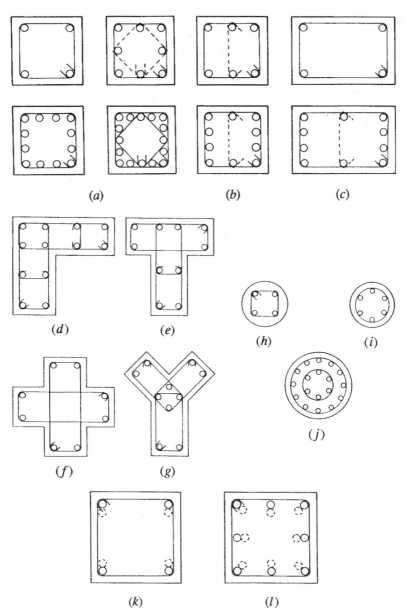

Figure 10.5 Considerations for bar layouts and tie patterns in tied concrete columns: (*a*) square columns with symmetrical reinforcement, (*b*) square columns with unsymmetrical reinforcement, (*c*) oblong columns, (*d*)–(*g*) oddly shaped columns, (*h*)–(*j*) round columns with tied or spiral reinforcement, and (*k*), (*l*) bar placement at location of lapped spice of vertical bars.

vertical continuity (splicing) of the steel bars, any of which may be appropriate for a given situation:

1. Bars may be lapped the required distance for development of the compression splice. For bars of smaller dimension and lower yield strengths, this is usually the desired method.
2. Bars may have milled square-cut ends butted together with a grasping device to prevent separation in a horizontal direction.
3. Bars may be welded with full-penetration butt welds or by welding of the grasping device described for method 2.

The choice of splicing methods is basically a matter of cost comparison, but is also affected by the size of the bars, by the degree of concern for bar spacing in the column arrangement, and possibly by a need for some development of tension through the splice, if uplift or high magnitudes of moments exist. If lapped splicing is used, a problem that must be considered is the bar layout at the location of the splice, at which point there will be twice the usual number of bars. The lapped bars may be adjacent to each other, but the usual considerations for space between bars must be taken into account. If spacing is not critical, the arrangement shown in Figure 10.5k is usually chosen, with the spliced sets of bars next to each other at the tie perimeter. If spacing limits prevent the arrangement in Figure 10.5k, that shown in Figure 10.5l may be used, with the lapped sets in concentric patterns. The latter arrangement is used for spiral columns, where spacing is often critical.

Bending of steel bars involves the development of yield stress to achieve plastic deformation (the residual bend). As bars get larger in diameter, it is more difficult—and less feasible—to bend them. Also, as the yield stress increases, the bending effort gets larger. It is questionable to try to bend bars as large as No. 14 or No. 18 in any grade, and is also not advised to bend any bars with yield stress greater than 75 ksi [517 MPa]. Bar fabricators should be consulted for real limits of this nature.

10.7 COLUMNS IN SITECAST FRAMES

Reinforced concrete columns seldom occur as single, pin-ended members, as opposed to most wood columns and many steel columns. This condition may exist for some precast concrete columns, but almost all sitecast columns occur as members in frames, with interaction of the frame members in the manner of a so-called *rigid frame*.

Rigid frames derive their name from the joints between members, which are assumed to be *moment-resistive* (rotationally rigid), and thus capable of transmitting bending moments between the ends of the connected members. This condition may be visualized by considering the entire frame as being cut from a single piece of material, as shown in Figure 10.6a. The sitecast concrete frame and the all-welded steel frame most fully approximate this condition.

When the horizontal-spanning frame members (beams) are subjected to vertical gravity loads, the inclination of their ends to rotate transmits bending to the columns connected to their ends, as shown in Figure 10.6b. If the frame is subjected to lateral loads (often the case because rigid frames are frequently used for lateral bracing), the relative horizontal displacement of the column tops and bottoms (called *lateral drift*) transmits bending to the members connected to the columns (see Figure 10.6c). The combination of these loadings results in the general case of combined axial loads plus bending in all the members of a rigid frame.

Figure 10.6d shows the effect on the cross section of a column in a frame: a condition of axial compression plus bending. For some purposes, it is useful to visualize this as an analogous eccentric compressive force, with the bending produced by the product of the compression times its distance of eccentricity (dimension *e* in Figure 10.6d). It is, thus, possible to consider the column to have a maximum capacity for compression (with $e = 0$), which is steadily reduced as the eccentricity is increased. This is the concept of the interaction graph (see Figure 10.4).

10.8 MULTISTORY COLUMNS

Concrete columns occur frequently in multistory structures. In the sitecast structure, separate stories are typically cast in separate pours, with a cold joint (construction joint) between the successive pours. Although this makes for a form of discontinuity, it does not significantly reduce the effective monolithic nature of the framed structure. Compression is continuous by the simple stacking of the levels of the heavy structure, and splicing of the reinforcement develops a form of tension continuity, permitting development of bending moments.

The typical arrangement of reinforcement in multistory columns is shown in Figure 8.5, which illustrates the form of bar development required to achieve the splicing of the reinforcement. This is essentially compressive reinforcement, so its development is viewed in those terms. However, an important practical function of the column bars is simply to tie the structure together through the discontinuous construction joints.

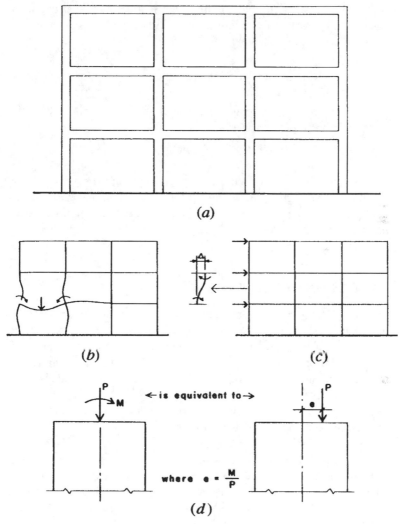

Figure 10.6 Columns in rigid frames.

Load conditions change in successive stories of the multistory structure. It is therefore common to change both the column size and reinforcement. Design considerations for this are discussed in the examples in Chapter 16.

In very tall structures, the magnitude of compression in lower stories requires columns with very high resistance. There is often some practical limit to column sizes, so that all efforts are made to obtain strength increases by means other than simply increasing the mass of concrete. The three basic means of achieving this are:

1. Increase the amount of reinforcement, packing columns with the maximum amount that is feasible and allowable by codes.
2. Increase the yield strength of the steel, using as much as twice the strength used for ordinary bars.
3. Increase the strength of the concrete.

The superstrength column is a clear case for use of the highest achievable concrete strengths, and is indeed the application that has resulted recently in spiraling high values for design strength. Strengths exceeding 20,000 psi have already been achieved, and higher ones are being proposed.

10.9   DESIGN METHODS AND AIDS

At present, design of concrete columns is mostly achieved by using either tabulations from handbooks or computer-aided procedures. Using the code formulas and requirements to design by "hand operation," with both axial compression and bending present at all times, is prohibitively laborious. The number of variables present (column shape and size, f'_c, f_y, number and size of bars, arrangement of bars, etc.) adds to the usual problems of column design, which makes for a situation much more complex than those for wood or steel columns.

The large number of variables also works against the efficiency of handbook tables. Even if a single concrete strength (f'_c) and a single steel yield strength (f_y) are used, tables would be very extensive if all sizes, shapes, and types (tied and spiral) of columns were included. Even with a very limited range of variables, handbook tables are much larger than those for wood or steel columns. They are, nevertheless, often quite useful for preliminary design estimation of column sizes. The obvious preference, when relationships are complex, requirements are tedious and extensive, and there are a large number of variables, is for a computer-aided system. It is hard to imagine a professional design office that is turning out designs of concrete structures on a regular basis, at the present time, without computer-aided methods. The reader should be aware that the software required for this work is readily available.

As in other situations, the common practices at any given time tend to narrow down to a limited usage of any type of construction, even though the potential for variation is extensive. It is thus possible to use some very limited but easy-to-use design aids to make early selections for design. These approximations may be adequate for preliminary building planning, cost estimates, and some preliminary structural analyses.

10.10　APPROXIMATE DESIGN OF TIED COLUMNS

Tied columns are much preferred to other columns due to the relative simplicity and usually lower cost of their construction, plus their adaptability to column shapes (square, round, oblong, T-shape, L-shape, etc.). Round columns—most naturally formed with spiral-bound reinforcing—are often made with ties instead, when the structural demands are modest.

The column with moment is often designed using the equivalent eccentric load method. The method consists of translating a compression-plus-bending situation into an equivalent one with an eccentric load, the moment becoming the product of the load and the eccentricity (see Figure 10.6d). This method is often used in presentation of tabular data for column capacities.

Figures 10.7 through 10.10 yield safe, ultimate factored capacities for a selected number of sizes of square tied columns with varying percentages of reinforcement. Allowable axial compression loads are given for various degrees of eccentricity, which is a means for handling axial load and bending moment combinations. The computed moment on the column is translated into an equivalent eccentric loading. Data for the curves were computed by strength design methods, as currently required by the ACI Code.

When bending moments are relatively high in comparison to axial loads, round or square column shapes are not the most efficient, just as they are not for spanning beams. Figures 10.11 and 10.12 yield safe, ultimate factored capacities for columns with rectangular cross sections. To further emphasize the importance of major bending resistance, all the reinforcement is assumed to be placed on the narrow sides, thus utilizing it for its maximum bending resistance effect.

The following examples illustrate the use of Figures 10.7 through 10.12 for the design of square and rectangular tied columns:

Example 1.　A square tied column with f'_c = 5 ksi [34.5 MPa] and steel with f_y = 60 ksi [414 MPa] sustains an axial compression load of 150

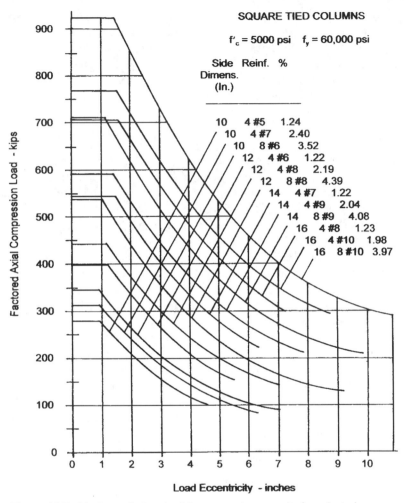

Figure 10.7 Maximum factored axial compression capacity for selected square tied columns.

kips [667 kN] dead load and 250 kips [1110 kN] live load, with no computed bending moment. Find the minimum practical column size if reinforcement is a maximum of 4% and the maximum size if reinforcement is a minimum of 1%.

Solution: As in all problems, this one begins with the determination of the factored ultimate axial load P_u

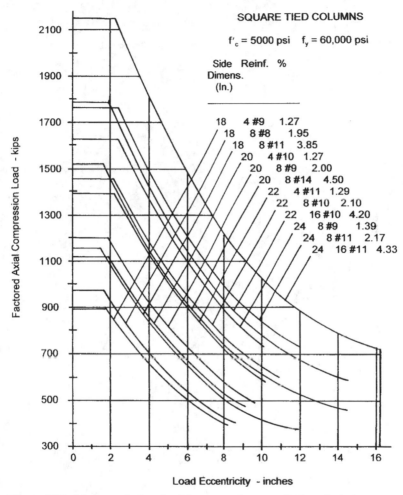

Figure 10.8 Maximum factored axial compression capacity for selected square tied columns.

$$P_u = 1.2P_{DL} + 1.6P_{LL} = (1.2 \times 150) + (1.6 \times 250) = 580 \text{ kips } [2580 \text{ kN}]$$

With no real consideration for bending, the maximum axial load capacity may be determined from the graphs by simply reading up the left edge of the figure. The curved lines actually end some distance from this edge, since the code requires a minimum bending for all columns.

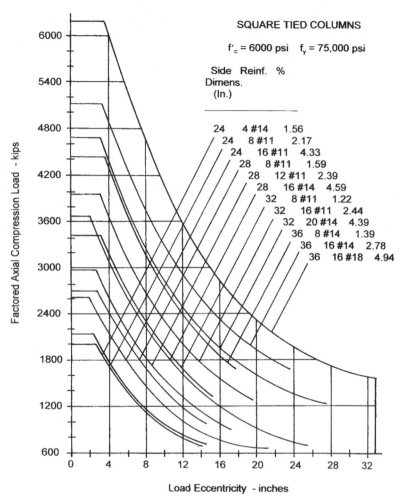

Figure 10.9 Maximum factored axial compression capacity for selected square tied columns.

Using Figure 10.7, the minimum size is a 14-in. square column with four No. 9 bars, for which the graph yields a maximum capacity of approximately 590 kips. Note that this column has a steel percentage of 2.04%.

What constitutes the maximum size is subject to some judgment. Any column with a curve above that of the chosen minimum column will work. It becomes a matter of increasing redundancy of capacity. How-

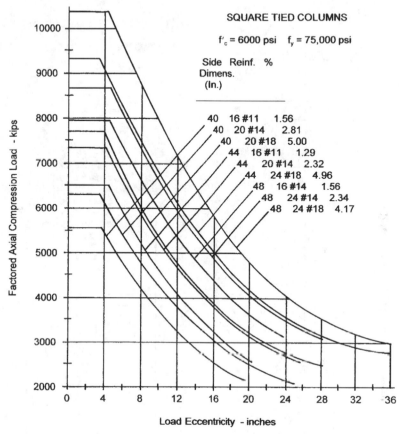

Figure 10.10 Maximum factored axial compression capacity for selected square tied columns.

ever, there are often other design considerations involved in developing a whole structural system, so these examples are quite academic. See the discussion for the building in Section 16.16. For this example, it may be observed that the minimum size choice is for a 14-in. square column. Thus, going up to a 15-in. or 16-in. size will reduce the reinforcement. We may thus note from the limited choices in Figure 10.7 that a maximum size is 16 in. square with four No. 8 bars, capacity is 705 kips, and $p_g = 1.23\%$. Because this is close to the usual recommended minimum reinforcement percentage (1%), columns of larger size will be increasingly redundant in strength (structurally oversized, in designer's lingo).

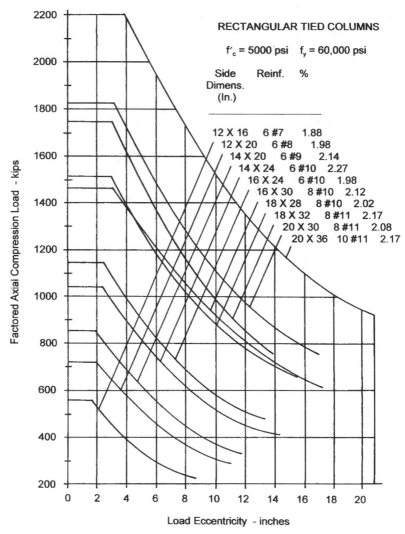

Figure 10.11 Maximum factored axial compression capacity for selected rectangular tied columns. Bending moment capacity determined for the major axis with reinforcement equally divided on the short sides of the section.

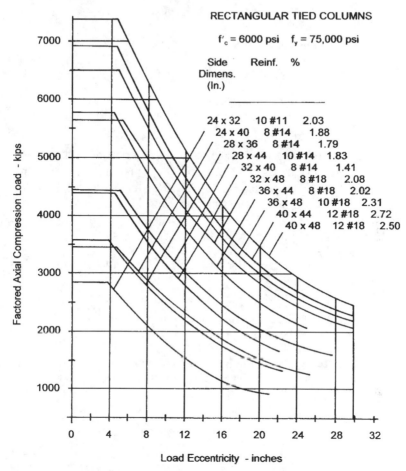

Figure 10.12 Maximum factored axial compression capacity for selected rectangular tied columns. Bending moment capacity determined for the major axis with reinforcement equally divided on the short sides of the section.

Example 2. A square tied column with $f'_c = 5$ ksi [34.5 MPa] and steel with $f_y = 60$ ksi [414 MPa] sustains an axial load of 150 kips [667 kN] dead load and 250 kips [1110 kN] live load, and a bending moment of 75 kip-ft [112 kN-m] dead load and 125 kip-ft [170 kN-m] live load. Determine the minimum-size column and its reinforcement.

Solution: First, determine the ultimate axial load and ultimate bending moment. From Example 1, $P_u = 580$ kips [2580 kN].

$$M_u = 1.2 \times M_{DL} + 1.6 \times M_{LL} = (1.2 \times 75) + (1.6 \times 125)$$

$$= 290 \text{ kip-ft [393 kN-m]}$$

Next, determine the equivalent eccentricity. Thus

$$e = \frac{M_u}{P_u} = \frac{290 \times 12}{580} = 6 \text{ in. [152 mm]}$$

Then, from Figure 10.8, minimum size is 18 in. square with eight No. 11 bars, and capacity at 6-in. eccentricity is approximately 650 kips. Note that the steel percentage is 3.85%. If this is considered to be too high, use a 20 × 20-in. column with four No. 10 bars with a capacity of approximately 675 kips, or a 22-by-22-in. column with four No. 11 bars with a capacity of approximately 900 kips.

Example 3. Select the minimum-size rectangular column for the same data as used in Example 2.

Solution: With the factored axial load of 580 kips and the eccentricity of 6 in., Figure 10.11 yields the following: 14-by-24-in. column, six No. 10 bars, with a capacity of approximately 730 kips.

Problems 10.10.A, B, C. Using Figures 10.7 through 10.10, select the minimum-size square tied column and its reinforcement for the following data.

TABLE 10.1   Data for Problem 10.10.A, B, C

	Concrete Strength (psi)	Axial Compressive Load (kips)		Bending Moment (kip-ft)	
		Live	Dead	Live	Dead
A	5000	80	100	30	25
B	5000	100	140	40	60
C	5000	150	200	100	100

Problems 10.10.D, E, F. From Figures 10.11 and 10.12, determine minimum sizes for rectangular columns for the same data as in Problems 10.10A, B, and C.

10.11   ROUND COLUMNS

Round columns, as discussed previously, may be designed and built as spiral columns, or they may be developed as tied columns with bars in a rectangular layout or with the bars placed in a circle and held by a series of round circumferential ties. Because of the cost of spirals, it is usually more economical to use the tied column, so it is often used unless the additional strength or other behavioral characteristics of the spiral column are required.

Figure 10.13 gives safe loads for round columns that are designed as tied columns. As for the square and rectangular columns in Figures 10.7 through 10.12, load values have been adapted from values determined by

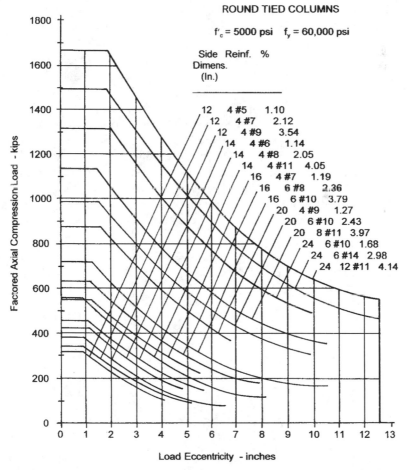

Figure 10.13 Maximum factored axial compression capacity for selected round tied columns.

strength design methods, and use is similar to that demonstrated in the preceding examples.

Problems 10.11.A, B, C.   Using Figure 10.13, pick the minimum-size round column and its reinforcing for the load and moment combinations in Problems 10.10.A, B, and C.

10.12 SPECIAL CONCERNS FOR CONCRETE COLUMNS

Slenderness

Cast-in-place concrete columns tend to be quite stout in profile, so that buckling failure related to slenderness is much less often a critical concern than with columns of wood or steel. Earlier editions of the ACI Code provided for consideration of slenderness, but permitted the issue to be neglected when the L/r of the column fell below a controlled value. For rectangular columns, this usually meant that the effect was ignored when the ratio of unsupported height-to-side dimension was less than about 12. This is roughly analogous to the case for the wood column with L/d less than 11.

Slenderness effects must also be related to the conditions of bending for the column. Because bending is usually induced at the column ends, the two typical cases are those shown in Figure 10.14. If a single end moment exists or two equal end moments exist as shown in Figure 10.14*a,* the buckling effect is magnified, and the *P*-delta effect is maximum. The condition in Figure 10.14*a* is not the common case, however; the more

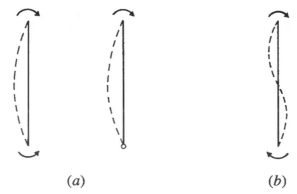

(a)   (b)

Figure 10.14   Assumed conditions of bending moments at column ends for consideration of column slenderness.

typical condition in framed structures is that shown in Figure 10.14*b,* for which the code treats the problem as one of moment magnification.

When slenderness must be considered, the ACI Code provides procedures for a reduction of column axial load capacity. One should be aware, however, that reduction for slenderness is not considered in design aids, such as tables or graphs.

Development of Compressive Reinforcement

In multistory buildings, it is usually necessary to splice the vertical reinforcement in columns. Steel bars are available in limited lengths, making it necessary to do some splicing between stories of the structure. Thus, the vertical load transfer from an upper column to the column below it is achieved in two parts: from concrete-to-concrete by direct bearing, and from steel-to-steel by splicing. This transfer must also occur at the joint between the lowest column and its supporting foundation. These development problems are treated in Section 8.5.

10.13   VERTICAL CONCRETE COMPRESSION ELEMENTS

Several types of construction elements are used to resist vertical compression for building structures. Dimensions of elements are used to differentiate between the defined elements. Figure 10.15 shows four such elements, described as follows:

- *Wall.* Walls of one or more story height are often used as bearing walls, especially in concrete and masonry construction. Walls may be quite extensive in length, but are also sometimes built in relatively short segments.
- *Pier.* When a segment of wall has a length that is less than six times the wall thickness, it is called a *pier,* or sometimes a *wall pier.*
- *Column.* Columns come in many shapes, but generally have some extent of height in relation to dimensions of the cross section. The usual limit for consideration as a column is a minimum height of three times the column diameter (side dimension, etc.). A wall pier may serve as a column, so the name distinction can be somewhat ambiguous.
- *Pedestal.* A *pedestal* is really a short column, that is, a column with height not greater than three times its thickness. This element is also frequently called a *pier,* adding to the confusion of names.

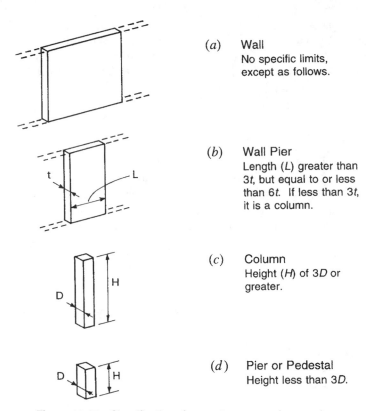

(*a*)   Wall
       No specific limits,
       except as follows.

(*b*)   Wall Pier
       Length (*L*) greater than
       3*t*, but equal to or less
       than 6*t*. If less than 3*t*,
       it is a column.

(*c*)   Column
       Height (*H*) of 3*D* or
       greater.

(*d*)   Pier or Pedestal
       Height less than 3*D*.

Figure 10.15   Classification of concrete compression members.

To add more confusion, most large, relatively stout and massive concrete support elements are typically also called *piers*. These may be used to support bridges, longspan roof structures, or any other extremely heavy load. Identity in this case is more a matter of overall size than any specific proportions of dimensions. Bridge supports and supports for arch-type structures are also sometimes called *abutments*.

One more use of the word "pier" is for description of a type of foundation element that is also sometimes called a *caisson*. This consists essentially of a concrete column cast in a vertical shaft that is dug in the ground.

Walls, piers, columns, and pedestals may also be formed from concrete masonry units (CMUs). The pedestal, as used with foundation systems, is discussed in Chapter 13.

10.14 CONCRETE MASONRY COLUMNS AND PIERS

Structural columns may be formed with concrete masonry construction for use as entities or as part of a general CMU structure. In light construction, column pedestals are commonly formed with CMU construction, especially if no other sitecast concrete is being used, other than for ground slabs and foundations.

Figure 10.16 shows several forms of CMU columns, commonly used with construction that is generally reinforced to qualify as structural ma-

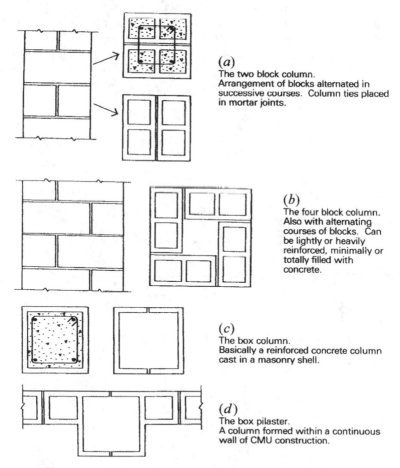

(*a*)
The two block column.
Arrangement of blocks alternated in successive courses. Column ties placed in mortar joints.

(*b*)
The four block column.
Also with alternating courses of blocks. Can be lightly or heavily reinforced, minimally or totally filled with concrete.

(*c*)
The box column.
Basically a reinforced concrete column cast in a masonry shell.

(*d*)
The box pilaster.
A column formed within a continuous wall of CMU construction.

Figure 10.16 Forms of CMU (concrete masonry unit) columns.

sonry. Figure 10.16*a* shows the minimum column, formed with two block units in plan. The positions of the two blocks are ordinarily rotated 90° in alternating courses of the construction, as shown in the figure. This column is totally filled with concrete and ordinarily reinforced with a vertical rod in each block cavity. Horizontal ties—necessary for full qualification as a structural column—must be placed in the mortar joints, which must usually be at least one-half-inch thick to accommodate the ties.

Figure 10.16*b* shows the four-unit column, forming in this case a small void area in the center of the column. Capacity of this column may be varied, with the minimum column having concrete fill and steel rods in only the corner voids. Additional rods and fill may be placed in the other voids for a stronger column. Finally, the center void may also be filled.

Even larger columns may be formed with a perimeter of CMU construction and an increasing center void. These may be constituted as hollow masonry shell structures, or may have significantly large concrete columns or piers cast into their voids.

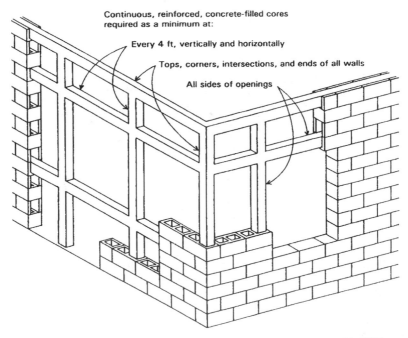

Continuous, reinforced, concrete-filled cores required as a minimum at:

Every 4 ft, vertically and horizontally

Tops, corners, intersections, and ends of all walls

All sides of openings

Figure 10.17 Common form of reinforced masonry construction with CMUs.

It is also possible to form a reinforced concrete column by casting the concrete inside a boxlike shell made from CMU pieces that define a considerable void. The simplest form for this is shown in Figure 10.16c, using two U-shaped units for each course of the masonry. Columns as small as 8 inches wide could be made this way, but the usual smallest size is one with a 12-inch-wide side, and the most common size is one with a 16-inch side, producing an exterior that exactly resembles the column in Figure 10.16a. With 16-inch units, the net size of the concrete column on the inside is about 13.5 inches, which is a significant concrete column.

The form of column in Figure 10.16c is also frequently used to produce pilasters in continuous walls of CMU construction. This is typically done by using alternating courses of units, with one course being as shown in Figure 10.16c and the alternating course being as shown in Figure 10.16d.

Shown in Figure 10.17 is a common form of structural masonry with CMUs, called *reinforced masonry*. In this type of construction, concrete is used in two ways:

- For precast units that are laid up with mortar in the time-honored fashion
- To fill selected vertically aligned voids and special horizontal courses after steel rods have been inserted

The result of using the concrete fill and steel reinforcement is a reinforced concrete rigid frame inside the CMU construction. This type of construction is most popular in western and southern states in the U.S.

11

COLUMN AND BEAM FRAMES

A common form of building structure, historically developed in wood and steel before concrete, is the column and beam frame. This system uses widely spaced columns for the major vertical support subsystem and horizontal beam networks for the roof and floor support subsystems. In response to the need to generate structures for multistory buildings, certain basic forms of this system were developed in wood and steel more than 100 years ago; these are mostly still in use, with little basic variation.

This chapter deals with various considerations for design of multistory column and beam systems using reinforced concrete. In wood and steel structures, the individual frame members are mostly all individual pieces, making their connections a major issue for design consideration. Analogous to this is the use of precast elements of concrete, which are indeed used for many structures. However, the more common form of construction is still that using sitecast concrete, which is the major focus of discussions in this chapter.

Much material that relates to this topic is developed in the other chapters in this book that treat the subjects of beams, columns, foundations, and whole system development. References to these discussions are made here, the specific concentration in this chapter being the concerns for the column and beam frames.

11.1 TWO-DIMENSIONAL FRAMES

In most building structures that use column and beam frames, there is some reasonable order to the layout of the system. Columns ordinarily occur in rows, with even spacing in a row and regular spacing of the rows; thus, the beams that connect columns are also in regularly spaced order. In this situation, what occurs typically are vertical planes of columns and beams that define individually a series of two-dimensional bents. (See Figure 11.1.) These do not exist as total, freestanding entities, but do form a subset within the whole structure. Actions of these individual bents may be investigated for effects of both gravity and lateral loads.

In sitecast concrete construction, individual two-dimensional column and beam bents constitute rigid frames. The continuity of steel reinforcement and the casting operation ensure this behavior. Some aspects of this behavior, as it affects the columns, are discussed in Section 10.7, and the general structural nature of frames is discussed in Section 4.6. The basic

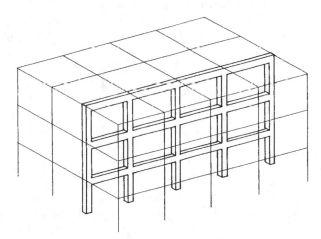

Figure 11.1 Two-dimensional column and beam bent as a subset in a three-dimensional framing system.

consideration is that both gravity and lateral loadings generate bending moments and shears in the columns.

For the beams in the sitecast frame, the natural form is that of horizontal continuity—for as many spans as are defined by the spaced columns in a single row. Thus, investigation and design of the beams for bending, shear, and development of reinforcement (Chapters 6, 7, and 8 respectively) must consider the continuous beam actions, as discussed in Section 9.1.

In addition to the actions of individual columns and beams, there is the action of the whole frame. This involves the collective *interactions* of all the columns and beams. One aspect of this is the highly statically indeterminate nature of the investigation for internal forces. In large frames, the full consideration for this behavior, including that for possible variations of loading, is itself a formidable computational problem, preceding any design efforts. Add to it the general complexity of design for reinforced concrete, and the design of the sitecast column and beam frame becomes a major undertaking.

A primary use of the vertical rigid frame bent is that of bracing for lateral forces due to wind and earthquakes. Once constituted as a rigid (moment-resisting) frame, however, its continuous, indeterminate responses will occur for all loadings. Thus, both the individual responses to gravity and lateral loadings must be considered, as well as their potential combined effects.

The complete engineering investigation and design of sitecast frames is well beyond the scope of this book. Some of the general considerations for design of multistory frames and examples of methods for approximate design are treated in the discussion of the Building Five example in Chapter 16.

11.2 THREE-DIMENSIONAL FRAMES

Although it is possible to visualize the individual two-dimensional (2D) bents in a sitecast column and beam frame, as shown in Figure 11.1, building structures are necessarily three-dimensional (3D) in form. Thus, the sitecast framework is indeed of a 3D form with connected 2D bents. The individual 2D bents occur in parallel sets and intersect at right angles to form the 3D frame. Thus, the complexity of actions in individual 2D bents is dwarfed by comparison with the complexity of behavior of the full, 3D framework.

Design of sitecast frameworks is typically broken down, for practical purposes, to the design of individual bents, individual continuous beams, and individual multistory columns. Thus, each member—beam or column—of the framework is individually defined for construction purposes. However, some aspects of the relationships between the individual frame members must be considered, both for their structural behaviors due to the frame interactions and for their construction details. The following are some of these considerations:

Continuity of Beams. Horizontal reinforcement for individual beams is extended into the supporting columns, and sometimes through columns and into adjacent beams. Thus, the reinforcement in the top of one beam at its support, provided for negative bending moment, is extended to become part of the reinforcement for the beam on the other side of the shared support. Where double reinforcing is desired (for compression reinforcement), the bottom bars in either beam may be extended through the support to develop the compression reinforcement. Because the tops of all beams are typically aligned (at the top surface of the slab), the negative reinforcing bars in beams that intersect at right angles in plan would naturally occur in the same horizontal level; some adjustment must be made for this in one beam or the other.

Continuity of Columns. Columns in multistory structures are stacked on top of each other. The concrete of an upper column may, thus, simply bear on top of the column beneath it, like two bricks in a wall. However, the reinforcing bars must be extended (up or down) to develop their shared stress. (See discussion in Section 8.5.) This requires some cooperative design of the vertically adjacent columns, regarding the shape and size of the concrete member and the amount and arrangement of the reinforcement.

Traffic Jam of Reinforcement. With the continuity of the column reinforcement, and the continuity of the beam reinforcement in two directions, there is a three-way traffic jam of steel bars that can easily become a real problem. With all of the intersecting members heavily reinforced, this may have to be worked out with models or actual mock-ups. And good luck with trying to pour concrete into the melee! No room at the inn. Add to this that the columns have closely spaced ties or spirals and the beams have closely spaced stirrups, concentrated for the high end shears in the members.

Shared Response of Parallel 2D Bents. For lateral loading, the horizontal sitecast roof or floor slabs will typically be very stiff in their horizontal diaphragm actions. Thus, all 2D bents in a given direction will be deflected sideways the same amount at each story. Like a set of simultaneously compressed springs, they will each take a share of the total deflecting force in proportion to the stiffness of the individual bent. This must be considered for design; it may, in fact, be controlled by deliberately stiffening selected, individual bents.

These and some other considerations for the 3D framework are discussed for Building Five in Chapter 16.

The inherent capacity of sitecast concrete frames for rigid frame action makes them useful for lateral bracing. Producing the same actions in wood or steel frames requires a modification of the usual connections, to produce moment-resisting joints between members. In wood frames, this is seldom achievable for transfer of major bending moments. In steel frames, it is commonly done, but does require elaborate, time-consuming, and expensive development of special welded or high-strength-bolted connections. On the other hand, if the natural moment-resisting connections are not desired for some reason, it is more difficult to decouple the naturally continuous, sitecast frame than it is to develop moment-resisting connections in wood or steel.

11.3 MIXED FRAME AND WALL SYSTEMS

Most buildings consist of a mixture of framed systems and walls. For structural use, walls may vary in potential. Metal and glass skins on buildings are typically not components of the general building structure, even though they must have some structural character to resist gravity and wind effects. Walls of cast concrete or concrete masonry unit (CMU) construction are frequently used as parts of the building structure, which makes it necessary to analyze the relationships between the walls and the framed structure. This section treats some of the issues involved in this analysis.

Coexisting, Independent Elements

Frames and walls may act independently for some functions, even though they interact for other purposes. For low-rise buildings, walls are often used to brace the building for lateral forces, even when a complete gravity-load-carrying frame structure exists. Such is the typical case with light wood frame construction using plywood shear walls. It may also be the case for a concrete frame structure with cast concrete or concrete masonry walls.

Attachment of walls and frames must be done to ensure the actions desired. Walls are typically very stiff, while frames often have significant deformation due to bending in the frame members. If interaction of the walls and frames is desired, they may be rigidly attached to achieve the necessary load transfers. However, if independent actions are desired, it may be necessary to develop special attachments that achieve load transfer for some purposes while allowing for independent movements due to other effects. In some cases, total separation may be desired.

A frame may be designed for gravity load resistance only, with lateral load resistance developed by walls acting as shear walls (see Figure 11.2). This method usually requires that some elements of the frame function as collectors, stiffeners, shear wall end members, or chords for horizontal diaphragms. If the walls are intended to be used strictly for lateral bracing, care must be exercised in developing attachment of wall tops to overhead beams; deflection of beams must be permitted to occur without transferring loads to the walls below.

Load Sharing

When walls are rigidly attached to columns, they usually provide continuous lateral bracing in the plane of the wall. This permits the column to be designed only for the relative slenderness in the direction perpendicular to the wall. This is more often useful for wood and steel columns (for example, wood stud two-by-fours and steel W-shapes with narrow flanges), but it may be significant for a concrete or masonry column with a cross section other than square or round.

In some buildings, both walls and frames may be used for lateral load resistance at different locations or in different directions. Figure 11.3 shows four such situations. In Figure 11.3a, a shear wall is used at one end of the building and a parallel frame at the other end for the wind from one direction. These two elements will essentially share equally in the load distribution, regardless of their relative stiffness.

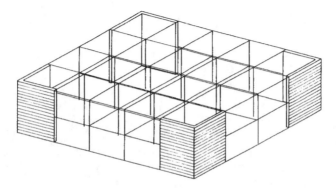

Figure 11.2   Framed structure braced by shear walls.

In Figure 11.3*b*, walls are used for the lateral loads from one direction and frames for the load in the perpendicular direction. Although some distribution must be made among the walls in one direction and among the frames in the other direction, there is essentially no interaction between the frames and walls, unless there is some significant torsion on the buildings as a whole.

Figures 11.3*c* and *d* show situations in which walls and frames do interact to share loads. In this case, the walls and frames share the total load from a single direction. If the horizontal structure is reasonably stiff in its own plane, the load sharing will be on the basis of the relative stiffnesses

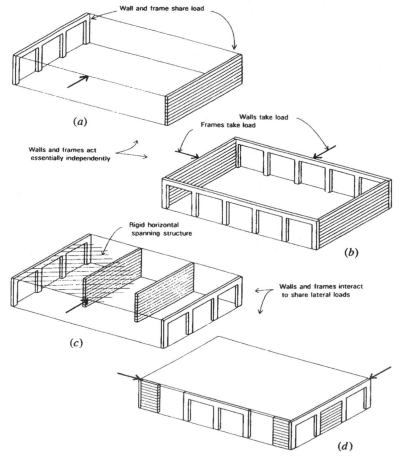

Figure 11.3   Lateral bracing systems with mixed shear walls and rigid frames.

of the vertical elements. Relative stiffness in this case refers essentially to resistance to deflection under lateral force.

Dual Systems

A dual system for lateral bracing is one in which a shear wall system is made to deliberately share loads with a frame system. In Figure 11.3, the systems shown at *a* and *b* are not dual systems, while those shown at *c* and *d* potentially are. The dual system has many advantages for structural performance, but the construction must be carefully designed and detailed to ensure that interactions and deformations do not result in excessive damage to the general construction. Some special problems that can occur are discussed in the next section.

11.4 SPECIAL PROBLEMS OF CONCRETE FRAMED BENTS

The rigid frame bent is a natural occurrence in ordinary sitecast construction with concrete columns and horizontal beam systems. This offers some potential advantages and some possible problems. Some of the design considerations were discussed in the preceding section; here, some additional concerns are addressed.

Inherent Lateral Load Capacity

Because of the continuity of the cast concrete and the steel reinforcement, rigid joints are largely unavoidable; thus, the sitecast concrete frame is naturally constituted as a rigid (moment-resistive) frame. For resistance to lateral forces, there is a free ride of a limited nature. That is, a certain level of moments in the joints and members is permitted with no additional consideration beyond normal design for gravity loads. Reasons for this are as follows:

Natural Form of Joints and Members.   Because of the continuity of the frame, the columns and beams will already be designed for some bending in the ends of the members and for some moment transfer through the joints due to gravity loads. Bending due to lateral forces is, thus, not a singular consideration for bending, but merely an additional one. The frame is already constituted to develop bending.

Modified Stress or Load Factors.   When forces due to wind or earthquakes are considered, an adjustment is made in allowable stresses (stress method) or in load factors (strength method). Thus, the addition of low levels of additional bending due to lateral forces may actually not require any additional member or joint capacity. That is to say, a certain amount of lateral force capacity is free of charge.

Minimum Bending in Columns. Current design requirements do not permit the design of a column for axial load only. A minimum level of bending (or minimum eccentricity of the load) is required. Thus, even without any bending induced by continuity of the beams, there is still some bending reserve in the columns.

Relative Stiffness of Parallel Bents

This was discussed in the preceding section. Bents that share lateral load in a single direction will each receive a portion of the load in proportion to the bent stiffnesses. Every bent will take some load, and if they are all equally stiff, they will each take an equal share. However, the ordinary case is one with bents of varying stiffnesses. Thus, it may be necessary to do deflection analyses of the parallel bents to determine the proportionate distribution.

If it is desired to assign lateral bracing to selected bents, a way to achieve the goal is simply to increase their stiffness. This is naturally achieved by increasing the relative stiffnesses of the bent members. Where heavier loads are carried by some columns or column-line beams, this may occur in the design for gravity loads. However, the bents so defined may not be the ones selected for lateral bracing. This issue is discussed in the development of the building design example in Section 16.16.

Proportionate Stiffness of Individual Bent Members

Within a single bent, the behavior of the bent and the forces in individual members will be strongly affected by the proportionate stiffness of bent members. If story heights vary and beam spans vary, some very complex and unusual behaviors may be involved. Variations of column and beam stiffnesses may also be a significant factor.

A particular concern is the relative stiffnesses of all the columns in a single story of the bent. In many cases, the portion of lateral shear in the columns will be distributed on this basis. Thus, the stiffer columns may carry a major part of the lateral force.

Another concern has to do with the relative stiffnesses of columns in comparison to beams. Most bent analyses assume the column stiffness to

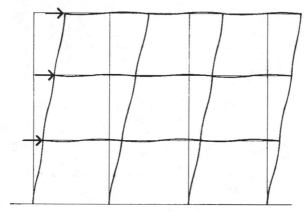

Figure 11.4 Deformation of a multistory rigid frame under lateral loading.

be more-or-less equal to the beam stiffness, producing the classic form of lateral deformation shown in Figure 11.4. Individual bent members are assumed to take an S-shaped, inflected form. However, if the columns are exceptionally stiff in relation to the beams, the form of bent deformation may be more like that shown in Figure 11.5a, with almost no inflection in the columns and an excessive deformation in upper beams. In tall frames, this is often the case in lower stories, where gravity loads require large columns.

Conversely, if the beams are exceptionally stiff in comparison to the columns, the form of bent deformation may be more like that shown in

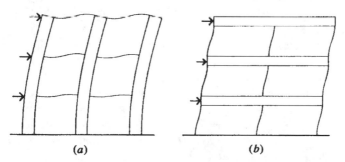

(*a*)   (*b*)

Figure 11.5 Form of deformation under lateral loading for rigid frames with members of disproportionate stiffness: (*a*) stiff columns and flexible beams, and (*b*) stiff beams and flexible columns.

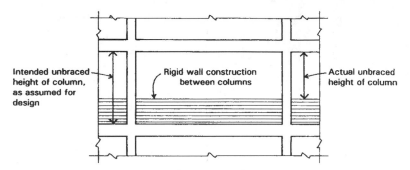

Figure 11.6   Common example of a captive column.

Figure 11.5*b,* with columns behaving as if fully fixed at their ends. Deep spandrel beams with relatively small columns commonly produce this situation.

All of the cases shown in Figures 11.4 and 11.5 can be dealt with for design, although it is important to understand which form of deformation is most likely.

The Captive Frame

In the preceding section, the problem of interaction of parallel bents and walls was discussed. A special problem is that of the partially restrained column or beam, with inserted construction that alters the form of deformation of bent members. An example of this, as shown in Figure 11.6, is the *captive column.* In the example, a partial-height wall is placed between columns. If this wall has sufficient stiffness and strength and is tightly wedged between columns, the laterally unbraced height of the column is drastically altered. As a result, the shear and bending in the column will be considerably different from that of the free column. In addition, the distribution of forces in the bent containing the captive columns may also be affected. Finally, the bent may, thus, be significantly stiffened, and its share of the load in relation to other parallel bents may be much higher.

This is an issue for the structural designer of the bents, but it must also be considered in cooperation with whomever does the construction detailing for the wall construction. This has been a major source of problems for concrete frames affected by seismic forces.

12

CONCRETE WALLS

Concrete is used extensively for walls, especially for subgrade construction. The three predominant forms for structural walls are sitecast concrete, precast concrete, and concrete masonry units (CMUs).

12.1 SITECAST WALLS: GENERAL CONCERNS

Concrete walls serve a variety of purposes in building construction. The following walls are common:

Bearing Walls, Uniformly Loaded.   Single-story or multistory, these carry loads from floors, roofs, and walls above.

Bearing Walls with Concentrated Loads.   These provide support for beams or columns. In most cases, they also support uniformly distributed loads.

Basement Walls, Earth-Retaining. These occur at the boundary between interior subgrade spaces and the surrounding earth. They not only act as bearing walls, but as spanning walls in resisting the horizontal pressure of the soil.

Retaining Walls. These achieve grade-level changes of site surfaces, working as vertical cantilevers to resist the horizontal earth pressures from the high side of the wall.

Shear Walls. These are used to brace buildings against wind and seismic forces. The shear is generated in the plane of the wall.

Freestanding Walls. Supported only at their bases, these serve as fences or partitions.

Grade Walls. These occur in buildings without basements, supporting walls, or columns above, as well as grade-level paving floor slabs. They may also serve as *grade beams* or *ties,* as described in Section 13.9.

Of course, walls can serve more than one function. Concrete walls are quite expensive, compared to other wall construction, so they are usually used for all the potential structural functions possible.

Concrete walls have two basic features as follows:

Wall Thickness. Nonstructural walls may be as thin as 4 in.; structural walls must be at least 6 in. thick. In general, the slenderness ratio (unbraced wall height divided by wall thickness) should not exceed 25. Usually, walls taller than 15 times their thickness require multiple pours. Walls greater than 10 in. thick should have two layers of reinforcement, one near each wall surface. Basement walls, foundation walls, and party walls must be at least 8 in. thick. Of course, a wall's thickness also depends on its structural tasks.

Reinforcement. Minimum areas of reinforcement are required, equal to 0.0025 times the wall cross section in a horizontal direction and 0.0015 in a vertical direction. This area may be reduced if No. 5 or smaller bars of Grade 60 or higher are used. As noted previously, walls greater than 10 in. thick require two layers of reinforcement; the distribution of the total area between the two layers depends on the wall's functions. Extra reinforcement should be provided at the wall top, bottom, ends, corners, and intersections, and around openings.

Regardless of its construction, every building wall has some specific architectural functions. Walls are often interrupted by openings for windows or doors, or to allow passage of ducts, piping, or wiring. Some of a wall's physical properties—such as fire rating, acoustic separation value, thermal insulative value, and so on—depend on its location and usage. Although concrete walls often serve structural functions, they must also be developed for the general purposes of the building.

12.2 CONCRETE BEARING WALLS

When the full wall cross section is utilized, bearing strength is limited as follows:

- By allowable stress method: $P = 0.30f'_c A_1$
- By strength method: $P_u = 0.7(0.85f'_c A_1)$

When the area developed in bearing is less than the total wall cross section, these loads may be increased by a factor equal to $\sqrt{A_2/A_1}$, but not more than 2. In these equations, A_1 is the actual area in bearing, and A_2 is the full wall cross section.

When the resultant vertical compression force on a wall falls within the middle third of the wall thickness, the wall may be designed as an axially loaded column, using the following empirical formula with the strength method:

$$\phi P_{nw} = 0.55\phi f'_c A_g \left[1 - \left(\frac{L_c}{32h} \right)^2 \right]$$

where
- ϕ = 0.70
- P_{nw} = nominal axial load strength of the wall
- A_g = effective area of the wall cross section
- L_c = vertical distance between lateral supports
- h = overall wall thickness

If the wall carries concentrated loads, the effective length of wall to be used for determination of A_g shall not exceed the center-to-center spacing of the loads nor the actual width of bearing plus four times the wall thickness.

The following example illustrates the design of a wall with concentrated loads. Designing for a uniformly distributed load is the same, except that bearing stress need not be considered.

Example 1. A reinforced concrete wall supports a roof system consisting of precast single Ts spaced 8 ft [2.44 m] on center. The stem of the Ts is 8 in. [203 mm] wide, but the bearing width is taken as 7 in. [178 mm] to allow for beveled bottom edges. The Ts bear on the wall's full thickness. The wall height is 11 ft 6 in. [3.51 m] and the reaction of each T due to service loads is 22 kips [98 kN] dead load and 12 kips [53.4 kN] live load. Using strength methods, design the wall given the following data: $f_c' = 4$ ksi [27.6 MPa], and $f_y = 40$ ksi [276 MPa]. (See Figure 12.1.)

Solution: The factored value of the reaction for a single T is

$$P_u = 1.2(\text{DL}) + 1.6(\text{LL}) = 1.2(22) + 1.6(12) = 45.6 \text{ kips [203 kN]}$$

And for the wall, the required resistance is

$$\frac{P_u}{\phi} = \frac{45.6}{0.7} = 65.1 \text{ kips [290 kN]}$$

Assume the minimum wall thickness of

$$h = \frac{L_c}{25} = \frac{11.5 \times 12}{25} = 5.52 \text{ in., say 6 in. [152 kN]}$$

For bearing, use the wall thickness, $h = 6$ in., and the bearing width, $b' = 7$ in. The bearing capacity is then

$$P_b = 0.85 f_c'(b' \times h) = 0.85(4)(7 \times 6) = 143 \text{ kips [636 kN]}$$

which indicates that bearing on the wall is not critical.

For column action, determine the effective horizontal length, which is controlled by the bearing width plus four times the wall thickness, or

$$b' + 4h = 7 + (4 \times 6) = 31 \text{ in. [787 mm]}$$

Check the L_c/h ratio:

$$\frac{L_c}{h} = \frac{11.5 \times 12}{6} = 23$$

Because this ratio does not exceed 25, the wall's capacity may be determined as

$$\frac{P_u}{\phi} = 0.55 f'_c A_g \left[1 - \left(\frac{L_c}{32h} \right)^2 \right]$$

$$= 0.55 \times 4 \times (6 \times 31) \left[1 - \left(\frac{11.5 \times 12}{32 \times 6} \right)^2 \right]$$

$$= 198 \text{ kips [881 kN]}$$

Because this is considerably larger than the required load of 65.1 kips, there is some margin left for possible eccentricity of the T reaction from the center of the wall.

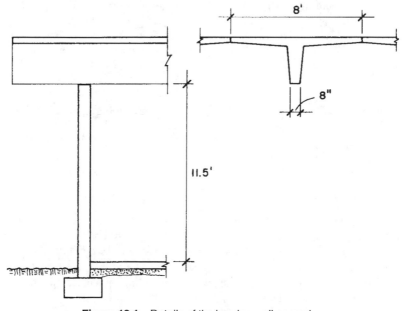

Figure 12.1   Details of the bearing wall example.

For the wall with vertical load only, reinforcement is usually adequate with satisfaction of the code limits for minimum reinforcement, expressed as a percentage of the wall gross-section area. Thus,

For vertical reinforcement:

$$A_s = 0.0015A_g = 0.0015(6 \times 12)$$

$$= 0.108 \text{ in.}^2 \text{ per linear ft } [229 \text{ mm}^2/\text{m}] \text{ of wall}$$

And for horizontal reinforcement:

$$A_s = 0.0025A_g = 0.0025(6 \times 12)$$

$$= 0.180 \text{ in.}^2 \text{ per linear ft } [381 \text{ mm}^2/\text{m}]$$

Maximum permitted bar spacing is three times the wall thickness, or 18 in. From Table 6.5, the vertical reinforcement may be No. 4 bars at 18 in., and the horizontal reinforcement may be No. 4 bars at 12 in.

Problem 12.2.A.   An 8-in.-thick [203 mm] concrete wall is 15 ft [4.6 m] high and supports precast concrete girders 10 ft [3.05 m] on center. Each girder has service reactions of 28 kips [125 kN] dead load and 14 kips [62 kN] live load. The girders have full bearing on the wall. The effective bearing width of the girder is 7.5 in. [191 mm]. Determine whether the wall is adequate for this loading, and select the required reinforcement. Use $f_c' = 3$ ksi [20.7 MPa] and $f_y = 40$ ksi [276 MPa].

Problem 12.2.B.   Same as Problem 12.1.A, except wall is 10 in. thick [254 mm], girders are 12 ft [3.66 m] on center, width of bearing is 9.5 in. [241 mm], DL = 36 kips [160 kN], LL = 18 kips [80 kN], $f_c' = 4$ ksi [27.6 MPa], and $f_y = 60$ ksi [414 MPa].

12.3   CONCRETE BASEMENT WALLS

A basement wall separates interior building space from earth, most of its height typically being below grade. Such walls should be properly treated for water resistance. They should also be reinforced for bending stresses due to the horizontal pressure of the earth.

Basement walls may be bearing walls, depending on a building's structural scheme. With respect to its earth-retaining function, a basement wall may serve as a spanning slab, either spanning vertically between the basement floor and the framed floor above, or horizontally between columns or intersecting walls.

Design for the horizontal earth pressure is usually done by assuming that the earth acts like a fluid, thus resulting in a pressure that varies from zero at the top (finish-grade level) to a maximum pressure at the bottom. At any given point on the wall, the magnitude of pressure is equal to the product of the weight of the fluid times the distance below the earth surface. See Figure 12.2a. In this case, the weight of the "fluid" is assumed to be some percentage of the weight of the earth materials—typically about one-third for a soil with mostly sand and gravel content. Building codes usually have data for determination of the pressure for various types of soil. For illustration purposes here, we will assume a horizontal pressure, q, equal to 35 lbs/sq ft per foot of depth below grade. For the height of the wall, it is common to use the symbol h, although this causes some confusion with the wall design in the preceding section, where h is used to designate the thickness of the wall.

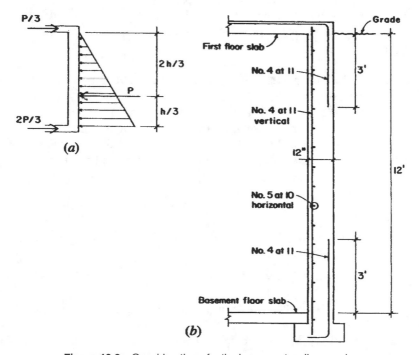

Figure 12.2   Considerations for the basement wall example.

With the pressure variation as shown in Figure 12.2*a,* the resultant total force on the wall, *P,* will act at the centroid of the pressure variation triangle, or one-third the distance from the bottom of the wall. The value of this force is determined as

$$P = \frac{qh^2}{2}$$

The reactions to the horizontal force will be one-third at the top of the wall and two-thirds at the bottom. These forces must be developed by the basement floor and the upper framed floor. The maximum bending moment in the wall for this triangular distribution of pressure is

$$M = 0.128 \, Ph$$

In many instances, the combination of required minimum wall thickness of 8 in. and the minimum percentages for vertical and horizontal reinforcement produces a wall that is adequate for relatively short walls with moderate bearing loads. Such is often the case for basements for residences.

Example 2. Using strength methods, design a reinforced concrete basement wall 12 ft [3.66 m] high between the basement and first floors. Exterior finished grade is approximately level with the bottom of the first-floor structure, as shown in Figure 12.2*b.* Assuming a minimal bearing load on the wall, design the wall using $f'_c = 3$ ksi [20.7 MPa] and $f_y = 60$ ksi [414 MPa]. Soil pressure is 35 psf/ft [0.51 kPa/m] of depth below grade.

Solution: Assuming a 1-ft-wide strip of wall (as in slab design), the total earth pressure is

$$P = \frac{qh^2}{2} = \frac{35 \times 12 \times 12}{2} = 2520 \text{ lb } [11.2 \text{ kN}]$$

The maximum moment in the wall is

$$M = 0.128Ph = 0.128(2520)(12) = 3871 \text{ ft-lb } [5.25 \text{ kN}]$$

This is a service load moment which is adjusted for strength design to

$$M_u = 1.6(3871) = 6194 \text{ ft-lb } [8.4 \text{ kN-m}]$$

The required nominal moment resistance for the wall is

$$M_r = \frac{M_u}{\phi} = \frac{6194}{0.9} = 6882 \text{ ft-lb } [9.33 \text{ kN-m}]$$

For the 12-in.-thick wall, the minimum vertical reinforcement is

$$A_s = 0.0015(12 \times 12) = 0.216 \text{ in.}^2 \text{ /ft } [457 \text{ mm}^2/\text{m}]$$

With this reinforcement placed ¾ in. from the inside wall surface, d will be approximately 11 in. With the very low percentage of steel area, a will be quite small, say 1 in. The moment capacity with the minimum reinforcement is

$$M = A_s f_y\left(d - \frac{a}{2}\right)$$

$$- (0.216)(60,000)(11 - 0.5)(1/12) = 11,340 \text{ ft-lb } [15.4 \text{ kN-m}]$$

Thus, the minimum reinforcement is adequate. This may be provided with No. 4 bars at 11 in. on center.

For horizontal reinforcement in the wall, a minimum percentage is 0.0025, in this case $0.0025(12 \times 12) = 0.36$ in.²/ft. This may be provided with No. 5 bars at 10-in. centers.

Depending on the forms for construction of the foundations and first floor, it may be advisable to provide for tension stresses in the wall at the top and bottom. Such reinforcement may be provided by bars near the outside surface of the wall, extending to the quarter height of the wall—in this case, equal to the vertical wall reinforcement. See Figure 12.2b.

Neither shear nor development length is likely to be critical in this wall, a common situation for basement walls of this type and size.

It is usually required that two layers of reinforcement be provided in walls of 12 in. thickness or greater, except for basement walls. While not required, many designers would choose to place two layers in this wall.

Problems 12.3.A, B. A basement wall is supported by the basement and first floor structures against the lateral earth pressure of 35 psf/ft [0.51 kPa/m] of depth below grade. Design the wall using $f_c' = 4$ ksi [27.6 MPa] and $f_y = 60$ ksi [414 MPa]. The wall height is: A, 15 ft [4.57 m]; B, 18 ft [5.49 m].

12.4 CONCRETE SHEAR WALLS

Because of their stiffness, dead weight, and potential strength, concrete walls are frequently used to brace buildings against the lateral forces of wind and earthquakes. Even when other bracing systems are present, concrete walls attract loading due to their high stiffness against shear forces in the plane of the wall.

Designing shear walls to resist either wind or earthquakes requires building planning, general lateral resistive system development, load determination, and investigation of connection and anchorage of system components. Some of these problems are treated in the design for the case examples in Chapter 16, but a complete treatment of the topic is beyond the scope of this book. For more extensive discussion, see *Simplified Building Design for Wind and Earthquake Forces* (Reference 6).

12.5 PRECAST CONCRETE WALLS

Many precast concrete walls are cast in a flat position at the building site. Even so, they are still classified essentially as *precast,* not *sitecast.* When the concrete is strong enough, the wall panels are lifted and placed in their desired positions for the building. This method is commonly known as *tilt-up construction,* which refers to its early development, when walls were cast immediately next to their desired location and then literally tilted up into position. Today such walls are lifted and placed with cranes, so that casting can make repeated use of single forms.

Some wall units are cast in factories, where precision of form, quality of materials, and finishing can be more highly controlled. Because of their large size and weight, these walls cannot be transported a great distance from the factory. Their use, is therefore, limited to a short range of distance from the precasting factory.

Precast units for roof and floor structures are usually prestressed. However, precast wall units are usually conventionally reinforced with inert steel bars. The major difference between precast and sitecast walls has to do with the necessary design for lifting and transporting effects for the precast units.

12.6 CONCRETE MASONRY WALLS

As discussed in Section 2.9, much of the structural masonry used for build-ing construction today is produced with units of precast concrete (concrete blocks, or as they are called now, CMUs, for *concrete masonry units*).

Unreinforced CMU construction is usually of the single-wythe (row) form shown in Figure 12.3a. The face shells, as well as the cross parts, are usually quite thick. This construction's structural integrity depends on the unit strength, the quality of the mortar, and the skill of the masons. Staggered vertical joints increase the bonding effect between adjacent units. Although it is called *unreinforced,* heavy wire elements are placed in horizontal joints, and vertical rods are inserted to strengthen wall ends and edges of openings.

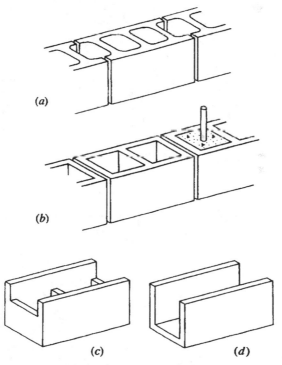

Figure 12.3 Forms of concrete masonry units (CMUs): (*a*) unit commonly used for unreinforced construction, (*b*) unit commonly used for reinforced construction, producing maximum size of interior voids, (*c*) unit used to create a horizontal reinforced beam within the solid wall, and (*d*) unit used for a lintel or header beam.

Reinforced CMU construction is produced with the units shown in Figure 12.3*b,* with large cavities and units stacked directly on top of each other to align the cavities vertically. When the cavities are filled with concrete and reinforced with vertical steel bars, a concrete column is formed inside the wall. Horizontal reinforcement goes in special horizontal courses produced with the modified block shown in Figure 12.3*c.* These units, as well as the one shown in Figure 12.3*d,* are also used to form beams to serve as lintels over openings in the wall.

Some examples of structural masonry are illustrated in Chapter 16. Masonry construction in general varies considerably by region, influenced by weather conditions and requirements of local building codes.

Reinforced structural masonry usually takes the form discussed in Sections 2.9 and 10.14 and shown in Figure 10.17.

13

FOUNDATIONS

Almost every building utilizes some concrete construction that is built directly in contact with the ground. Such elements may include the following:

- Shallow bearing footings, consisting of concrete pads that are used to spread the vertical loads of the building onto the supporting soil materials
- Concrete piles, pile caps, or concrete-filled excavated shafts used to develop deep foundations
- Concrete walls, enclosing below-grade spaces or forming grade beams
- Concrete pavements, for driveways, walks, parking lots, patios, or building floors placed directly on the ground
- Retaining walls, used to achieve sudden changes of the site surface profile

- Underground tunnels and vaults for service systems
- Bases for elevators and other equipment

This construction constitutes a major usage of concrete. Its low bulk cost, general nonrotting nature, rocklike character, general overall stiffness of thick elements, and ease of placement in the ground-level work make it the natural choice for most work. In days past, much of this construction was achieved with masonry materials. This is still possible, but now is generally limited to walls in special situations.

This is potentially a large topic if fully developed. Actually, there are two topics involved: concrete design and soil mechanics. The presentations in this chapter are limited to simple bearing-type foundation elements. Soil properties and usage are discussed briefly, as related to the bearing foundation situations. For a more extensive discussion of the topic, the reader is referred to *Simplified Design of Building Foundations* (Reference 7).

13.1 GENERAL CONCERNS FOR FOUNDATIONS

The design of the foundation for a building cannot be separated from the overall problems of the building structure and the building and site designs in general. Nevertheless, it is useful to consider the specific aspects of the foundation design that must be dealt with.

Site Exploration

For purposes of the foundation design, as well as for the building and site development in general, it is necessary to know the actual site conditions. This investigation usually consists of two parts: determination of the ground surface conditions and of the subsurface conditions. The surface conditions are determined by a site survey that establishes the three-dimensional geometry of the surface and the location of various objects and features on the site. Where they exist, the location of buried objects, such as sewer lines, underground power and telephone lines, and so on, may also be shown on the site survey.

Unless they are known from previous explorations, the subsurface conditions must be determined by penetrating the surface to obtain samples of materials at various levels below the surface. Inspection and testing of these samples in the field, and in a testing lab, are used to identify the materials and to establish a general description of the subsurface conditions.

Site Design

Site design consists of positioning the building on the site and the general development, or redevelopment, of the site contours and features. The building must be both horizontally and vertically located. Recontouring the site may involve both taking away existing material (called *cutting*) and building up to a new surface with materials brought in or borrowed from other locations on the site (called *filling*). Development of controlled site drainage for water runoff is an important part of the site design.

Selection of Foundation Type

The first formal part of the foundation design is the determination of the type of foundation system to be used. This decision cannot normally be made until the surface and subsurface conditions are known in some detail, and the general size, shape, and location of the building are determined. In some cases, it may be necessary to proceed with an approximate design of several possible foundation schemes so that the results can be compared.

Design of Foundation Elements

With the building and site designs reasonably established, the site conditions known, and the general type of foundation system determined, work can proceed to the detailed design of individual structural elements of the foundation system.

13.2   SOIL CONDITIONS RELATED TO FOUNDATION DESIGN

For the specific concerns of design and construction of foundations, some particular aspects of the soil conditions become most critical.

Structural Properties of Soil

The principal properties and behavior characteristics of soils that are of most direct concern in foundation design are the following:

Strength.   For bearing-type foundations, the main strength concern is for resistance to vertical compression. Resistance to horizontal pressure and to friction are of concern when foundations must resist horizontal forces due to wind, earthquakes, or retained soil.

Deformation Resistance. Deformation of soil under stress is of concern in designing for limitations of the movements of structures supported on soil, such as the vertical settlement of bearing foundations.

Stability. Various actions may produce changes in the physical character or form of soils; these include frost, fluctuations in water content, seismic shock, organic decomposition, and disturbances during construction. The degree of sensitivity of the soil to these actions is called its *relative stability.*

Properties Affecting Construction Activity

A number of possible factors relating to soil conditions may affect construction work, including the following:

* The relative ease of excavation
* Ease of performing and possible effects of dewatering during construction (lowering the ground water level)
* Feasibility of using excavated site materials as fill elsewhere on the site
* Ability of the unexcavated soil to stand on a steep cut face at the side of an excavation
* Effects of construction activity on unstable soils, notably the movement of workers and equipment on the soil surface

Miscellaneous Concerns

In specific situations, various factors may affect the design and construction work, including are the following:

* Location of the water table, affecting soil strength or stability, need for waterproofing for basements, need for dewatering during construction, and so on
* Nonuniform soil conditions, such as soil strata that are not horizontal, strips or pockets of poor soil, and so on
* Frost conditions, affecting the depth required for bearing foundations and possible heave and settlement of finish-grade surfaces and pavements
* Deep excavation or dewatering operations, possibly affecting stability of nearby properties, buildings, streets, and buried piping or tunnels

All of these concerns must be anticipated and dealt with in designing foundations, and a discovery plan for obtaining design information should be developed very early, so that any highly difficult situations are anticipated before considerable design work has been performed. Otherwise, extensive reworking of designs may be required.

13.3 FOUNDATION DESIGN: CRITERIA AND PROCESS

For the design of ordinary bearing-type foundations, values for several structural properties of a soil must be established. The principal values are the following:

Allowable Bearing Pressure.   This is the maximum permissible value for vertical compression stress at the contact surface of bearing elements. It is typically quoted in units of pounds or kips per square foot of contact surface.

Compressibility.   This is the predicted amount of volumetric consolidation that determines the amount of settlement of the foundation. Quantification is usually done in terms of the actual dimension of vertical settlement predicted for the foundation.

Active Lateral Pressure.   This is the horizontal pressure exerted against retaining structures, visualized in its simplest form as an equivalent fluid pressure. Quantification is in terms of a density for the equivalent fluid, given in actual unit weight value or as a percentage of the soil unit weight.

Passive Lateral Pressure.   This is the horizontal resistance offered by the soil to forces against the soil mass. It is also visualized as varying linearly with depth in the manner of a fluid pressure. Quantification is usually in terms of a specific pressure increase per unit of depth.

Friction Resistance.   This is the resistance to sliding along the contact-bearing face of a footing. For cohesionless soils (sands), it is usually given as a friction coefficient to be multiplied by the compressive force. For clays, it is given as a specific value in pounds per square foot to be multiplied by the contact area.

Whenever possible, stress limits should be established as the result of a thorough investigation and the recommendations of a qualified soils engineer. Most building codes allow for the use of so-called *presumptive*

TABLE 13.1 Presumptive Design Values for Soils[a]

Type of Soil	Allowable Foundation Pressure (psf)	Lateral Pressure (psf per ft of depth)	Sliding Resistance	
			Coefficient	Resistance (psf)
Bedrock, solid	4000	1200	0.70	
Bedrock, faulted	2000	400	0.35	
Gravel, sandy gravel	2000	200	0.35	
Sand, silty sand, clayey sand, silty gravel, and clayey gravel	1500	150	0.25	
Clay, sandy clay, silty clay, and clayey silt	1000	100		130

[a] Values to be used without testing of soils.

values for design, which may be used when soil investigation is not performed. Table 13.1 lists data of a form used in building codes for such values.

For very ordinary situations, building codes or code-enforcing agencies sometimes permit construction without submission of a detailed engineering design. Table 13.2 yields data for footings for simple stud-bearing wall construction of a light wood frame.

Figure 13.1 presents information for simple foundation elements for light wood frame structures.

13.4 SHALLOW BEARING FOUNDATIONS

The most common foundation consists of pads of concrete placed beneath the building. Because most buildings make a relatively shallow

TABLE 13.2 Foundations for Stud-Bearing Walls[a]

Number of Floors Supported by the Foundations	Thickness of Foundation Wall (in.)	Width of Footing (in.)	Thickness of Footing (in.)	Minimum Depth Below Undisturbed Ground Surface (in.)
1	6	12	6	12
2	8	15	7	18
3	10	18	8	24

[a] Where frost action is not critical.

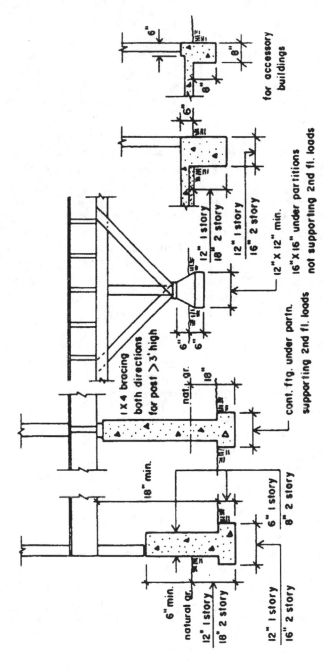

Figure 13.1 Foundations for light wood frame construction.

penetration into the ground, these pads, called *footings,* are generally classified as *shallow bearing foundations.* For simple economic reasons, shallow foundations are generally preferred. However, when adequate soil does not exist at a shallow location, driven piles or excavated piers (caissons)—which extend some distance below the building—must be used; these are called *deep foundations.*

The two common footings are the wall footing and the column footing. Wall footings occur in strip form, usually placed symmetrically beneath the supported wall. Column footings are most often simple square pads supporting a single column. When columns are very close together or at the very edge of the building site, special footings that carry more than a single column may be used.

Two other basic construction elements that occur frequently with foundation systems are foundation walls and pedestals. Foundation walls may be used as basement walls or merely as transition between more deeply placed footings and the aboveground building construction. Foundation walls are common with aboveground construction of wood or steel because these constructions must be kept from contact with the ground.

Pedestals are actually short columns used as transitions between the building columns and their bearing footings. These may also be used to keep wood or steel columns aboveground, or they may serve a structural purpose to facilitate the transfer of a highly concentrated force from a column to a widely spread footing.

13.5 WALL FOOTINGS

Wall footings consist of concrete strips placed under walls. The most common type is that shown in Figure 13.2, consisting of a strip with a rectangular cross section placed in a symmetrical position with respect to the wall, and projecting an equal distance as a cantilever from both faces of the wall. For soil pressure, the critical dimension of the footing is its width as measured perpendicular to the wall.

Wall footings ordinarily serve as construction platforms for the walls they support. Thus, a minimum width is established by the wall thickness plus a few inches on each side. The extra width is necessary because of the crude form of foundation construction, but it also may be required for support of forms for concrete walls. A minimum projection of 2 in. is recommended for masonry walls and 3 in. for concrete walls.

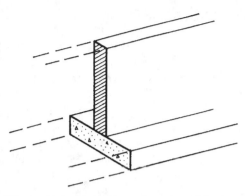

Figure 13.2 Typical form of a strip wall footing.

With relatively light vertical loads, the minimum construction width may be adequate for soil bearing. Walls ordinarily extend some distance below grade, and allowable bearing will usually be somewhat higher than for very shallow footings. With the minimum recommended footing width, the cantilever bending and shear will be negligible, so no transverse (perpendicular to the wall) reinforcement is used. However, some longitudinal reinforcement is recommended.

As the wall load increases and a wider footing is required, the transverse bending and shear require some reinforcement. At some point, the increased width also determines a required thickness. Otherwise, recommended minimum thickness is 8 in. for nonreinforced footings and 10 in. for reinforced footings.

Determination of the Footing Width

Footing width is determined by soil pressure, assuming that the minimum width required for construction is not adequate for bearing. Because footing weight is part of the total load on the soil, the required width cannot be precisely determined until the footing thickness is known. A common procedure is to assume a footing thickness, design for the total load, verify the structural adequacy of the thickness, and, if necessary, modify the width once the final thickness is determined. The current ACI Code (Reference 1) calls for using the unfactored loading (service load) when determining footing width by soil pressure.

Determination of the Footing Thickness

If the footing has no transverse reinforcement, the required thickness is determined by the tension stress limit of the concrete, in either flexural stress or diagonal stress due to shear. Transverse reinforcement is not required until the footing width exceeds the wall thickness by some significant amount, usually 2 ft or so. A good rule of thumb is to provide transverse reinforcement only if the cantilever edge distance for the footing (from the wall face to the footing edge) exceeds the footing thickness. For average conditions, this means transverse reinforcement for footings of about 3-ft width or greater.

If transverse reinforcement is used, the critical concerns become for shear in the concrete and tension stress in the reinforcing. Thicknesses determined by shear will usually ensure a low bending stress in the concrete, so the cantilever beam action will involve a very low percentage of steel. This is in keeping with the general rule for economy in foundation construction, which is to reduce the amount of reinforcement to a minimum.

Minimum footing thicknesses are a matter of design judgment, unless limited by building codes. The ACI Code recommends limits of 8 in. for unreinforced footings and 10 in. for footings with transverse reinforcement. Another possible consideration for the minimum footing thickness is the necessity for placing dowels for wall reinforcement.

Selection of Reinforcement

Transverse reinforcement is determined on the basis of flexural tension and development length due to the cantilever action. Longitudinal reinforcement is usually selected on the basis of providing minimum shrinkage reinforcement. A reasonable value for the latter is a minimum of 0.0015 times the gross concrete area (area of the cross section of the footing). Cover requirements are for 2 in. from formed edges and 3 in. from surfaces generated without forming (such as the footing bottom). For practical purposes, it may be desirable to coordinate the spacing of the footing transverse reinforcement with that of any dowels for wall reinforcement. Reinforcement for shear is required by code only when the ultimate shear capacity due to factored loading (V_u) is greater than the factored shear capacity of the concrete (ϕV_c).

The following example illustrates the design procedure for a reinforced wall footing. Data for predesigned footings are given in Table 13.3. Figure 13.3 provides an explanation of the table entries. The use of

TABLE 13.3   Allowable Loads on Wall Footings (see Figure 13.3)

Maximum Soil Pressure (lb/ft^2)	Minimum Wall Thickness, t (in.)		Allowable Load on Footing^a (lb/ft)	Footing Dimensions (in.)		Reinforcement	
	Concrete	Masonry		h	w	Long Direction	Short Direction
1000	4	8	2625	10	36	3 No. 4	No. 3 at 17
	4	8	3062	10	42	2 No. 5	No. 3 at 12
	6	12	3500	10	48	4 No. 4	No. 4 at 18
	6	12	3938	10	54	3 No. 5	No. 4 at 13
	6	12	4375	10	60	3 No. 5	No. 4 at 10
	6	12	4812	10	66	5 No. 4	No. 5 at 13
	6	12	5250	10	72	4 No. 5	No. 5 at 11
1500	4	8	4125	10	36	3 No. 4	No. 3 at 11
	4	8	4812	10	42	2 No. 5	No. 4 at 14
	6	12	5500	10	48	4 No. 4	No. 4 at 11
	6	12	6131	11	54	3 No. 5	No. 5 at 16
	6	12	6812	11	60	5 No. 4	No. 5 at 12
	6	12	7425	12	66	4 No. 5	No. 5 at 11
	8	16	8100	12	72	5 No. 5	No. 5 at 10
2000	4	8	5625	10	36	3 No. 4	No. 4 at 15
	6	12	6562	10	42	2 No. 5	No. 4 at 12
	6	12	7500	10	48	4 No. 4	No. 5 at 13
	6	12	8381	11	54	3 No. 5	No. 5 at 12
	6	12	9520	12	60	4 No. 5	No. 5 at 10
	8	16	10,106	13	66	4 No. 5	No. 5 at 10
	8	16	10,875	15	72	6 No. 5	No. 5 at 10
3000	6	12	8625	10	36	3 No. 4	No. 4 at 11
	6	12	10,019	11	42	4 No. 4	No. 5 at 14
	6	12	11,400	12	48	3 No. 5	No. 5 at 11
	6	12	12,712	14	54	6 No. 4	No. 5 at 11
	8	16	14,062	15	60	5 No. 5	No. 5 at 10
	8	16	15,400	16	66	5 No. 5	No. 6 at 13
	8	16	16,725	17	72	6 No. 5	No. 6 at 11

^{a}Allowable loads do not include the weight of the footing, which has been deducted from the total bearing capacity. Criteria: $f'_c = 2000$ psi, Grade 40 bars.

unreinforced footings is not recommended for footings greater than 3 ft in width.

Note: In using the strip method, a strip width of 12 in. is an obvious choice when using U.S. units, but not with metric units. To save space, the computations are performed with U.S. units only, but some metric equivalents are given for key data and answers.

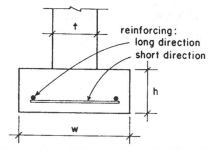

Figure 13.3 Reference figure for Table 13.3.

Example 1. Design a wall footing with transverse reinforcement for the following data:

- Footing design load = 3750 lb/ft [54.7 kN/m] dead load and 5000 lb/ft [73.0 kN/m] live load of wall length
- Wall thickness for design = 6 in. [150 mm]
- Maximum soil pressure = 2000 psf [96 kPa]
- Concrete design strength = 2000 psi [13.8 MPa]
- Steel yield stress = 40,000 psi [276 MPa]

Solution: For the reinforced footing, the only concrete stress of concern is that in shear. Concrete flexural stress will be low because of the low percentage of reinforcement. As with the unreinforced footing, the usual design procedure consists of making a guess for the footing thickness, determining the required width for soil pressure, and then checking the footing stress.

Try $h = 12$ in. Then footing weight = 150 psf, and the net usable soil pressure is $2000 - 150 = 1850$ lb/ft^2.

The footing design load is unfactored when determining footing width; therefore, the load is 8750 lb/ft [128 kN/m]. Required footing width is $8750/1850 = 4.73$ ft, or $4.73(12) = 56.8$ in., say 57 in., or 4 ft 9 in., or 4.75 ft [1.45 m]. With this width, the design soil pressure for stress is $8750/4.75 = 1842$ psf [88 kPa].

For the reinforced footing, it is necessary to determine the effective depth; that is, the distance from the top of the footing to the center of the steel bars. For a precise determination, this requires a second guess: the steel bar diameter (D). For the example, a guess is made of a No. 6 bar with a diameter of 0.75 in. With the cover of 3 in., this produces an effective depth of $d = h - 3 - (D/2) = 12 - 3 - (0.75/2) = 8.625$ in. [219 mm].

Concern for precision is academic in footing design, however, considering the crude nature of the construction. The footing bottom is formed by a hand-dug soil surface, unavoidably roughed-up during the placing of the reinforcement and casting of the concrete. The value of d will therefore be taken as 8.6 in. [218 mm].

Next, we need to determine how much the soil is pushing back up on the footing, using the factored loads. This load is

$$w_u = (1.2 \times 3750) + (1.6 \times 5000) = 12{,}500 \text{ lb/ft } [182 \text{ kN/m}]$$

With a width of 4.75 ft, the factored design soil pressure is

$$P_d = \frac{12{,}500}{4.75} = 2632 \text{ psf } [126 \text{ kPa}]$$

The critical section for shear stress is taken at a distance of d from the face of the wall. As shown in Figure 13.4a, this places the shear section at a distance of 16.9 in. from the footing edge.

At this location, the shear force is determined as

$$V_u = (2632 \text{ plf}) \times (16.9 \text{ in.}) \times \left(\frac{1 \text{ ft}}{12 \text{ in.}} \right) = 3707 \text{ lb } [16.5 \text{ kN}]$$

and the shear capacity of the concrete is

$$\phi V_c = 0.75 \, (2\sqrt{f'_c})(b \times d) = 0.75(2\sqrt{20000})(12 \times 8.6) = 6923 \text{ lb } [30.8 \text{ kN}]$$

It is possible, therefore, to reduce the footing thickness. However, cost effectiveness is usually achieved by reducing the steel reinforcement to a minimum. Low-grade concrete dumped into a hole in the ground is quite inexpensive, compared to the cost of steel bars. Selection of footing thickness, therefore, becomes a matter of design judgment, unless the footing width becomes as much as five times or so the wall thickness, at which point concrete stress limits may become significant.

If a thickness of 11 in. is chosen for this example, the shear capacity decreases only slightly, and the required footing width will remain effec-

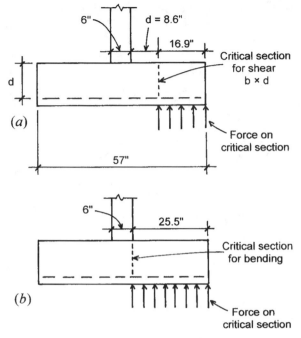

Figure 13.4 Shear and bending considerations for the wall footing.

tively the same. A new effective depth of 7.6 in. will be used, but the design soil pressure for stresses will remain the same because it relates only to the width of the footing.

The bending moment to be used for determination of the steel bars is computed as follows (see Figure 13.4*b*):

The force on the cantilevered edge of the footing is

$$F = \frac{25.5}{12} \times 2632 = 5593 \text{ lb [24.9 kN]}$$

and the cantilever bending moment at the wall face is, thus,

$$M_u = 5593 \times \frac{25.5}{2} = 71{,}310 \text{ in.-lb} = 5942 \text{ ft.-lb [8.06 kN-m]}$$

$$M_t = \frac{M_u}{\phi} = \frac{71,310}{0.9} = 79,233 \text{ in-lb.} = 6603 \text{ ft-lb } [8.95 \text{ kN-m}]$$

and the required steel area per foot of wall length is

$$A_s = \frac{M_t}{f_y(d - \frac{a}{2})} = \frac{79,233}{40,000 \times 0.9 \times 7.6} = 0.290 \text{ in.}^2 [187 \text{ mm}^2]$$

The spacing required for a given bar size to satisfy this requirement can be derived as follows:

$$\text{Required spacing} = (\text{area of bar}) \times \frac{12}{\text{required area/ft}}$$

Thus, for a No. 3 bar

$$s = 0.11 \times \frac{12}{0.290} = 4.6 \text{ in. } [117 \text{ mm}]$$

In this procedure, the required spacings for bar sizes 3 through 7 are shown in the fourth column of Table 13.4.

Bar sizes and spacings can be most easily selected using handbook tables that yield the average steel areas for various combinations of bar size and spacing. One such table is Table 6.5, from which the spacing figures shown in the last column of Table 13.4 were selected, indicating a range of choices for the footing transverse reinforcement. Selection of the ac-

TABLE 13.4 Selection of Reinforcement for Example 1

Bar Size	Area of Bar (in.²)	Area Required for Bending (in.²)	Bar Spacing Required (in.)	Bar Spacing Selected (in.)
3	0.11	0.290	4.6	4.5
4	0.20	0.290	8.3	8.0
5	0.31	0.290	12.8	12.5
6	0.60	0.290	24.8	18.0

248   FOUNDATIONS

tual bar size and spacing is a matter of design judgment, for which some considerations are as follows:

1. Maximum recommended spacing is 18 in.
2. Minimum recommended spacing is 6 in. to reduce the number of bars and make placing of concrete easier.
3. Preference is for smaller bars, as long as spacing is not too close.
4. A practical spacing may be that of the spacing of vertical reinforcement in the supported wall, for which footing dowels are required (or some full-number multiple or division of the wall bar spacing).

With these considerations in mind, a choice may be made for either the No. 5 bars at 12.5-in. spacing or the No. 4 bars at 8-in. spacing. The No. 6 bars at 24-in. spacing would not be a good choice due to the large distance between bars; thus, the practice is to use a maximum spacing of 18 in. for any spaced set of bars. Another consideration that must be made for the choice of reinforcement is the required development length for anchorage. With 2 in. of edge cover, the bars will extend 23.5 in. from the critical bending section at the wall face (see Figure 13.4*b*). Inspection of Table 8.1 will show that this is an adequate length for all the bar sizes used in Table 13.4. Note that the placement of the bars in the footing falls in the classification of "other" bars in Table 8.1.

For the longitudinal reinforcement, the minimum steel area is

$$A_s = (0.0015)(11)(57) = 0.94 \text{ in.}^2 \text{ [606 mm}^2\text{]}$$

Using three No. 5 bars yields

$$A_s = (3)(0.31) = 0.93 \text{ in.}^2 \text{ [600 mm}^2\text{]}$$

Table 13.3 gives values for wall footings for four different soil pressures. Table data were derived using the procedures illustrated in the example. Figure 13.3 shows the dimensions referred to in the table.

Problem 13.5.A.   Using concrete with a design strength of 2000 psi [13.8 MPa] and Grade 40 bars with a yield strength of 40 ksi [276 MPa], design a wall footing for the following data: wall thickness = 10 in. [254 mm], dead load on footing = 5000 lb/ft [73 kN/m], and live load = 7000 lb/ft [102 kN/m]; maximum soil pressure = 2000 psf [96 kN/m²].

Problem 13.5.B.   Same as Problem 13.5.A, except wall is 15 in. [380 mm] thick, dead load is 6000 lb/ft [87.5 kN/m], and live load is 8000 lb/ft [117 kN/m]; maximum soil pressure is 3000 psf [144 kN/m^2].

13.6   COLUMN FOOTINGS

The great majority of independent or isolated column footings are square in plan, with reinforcement consisting of two equal sets of bars at right angles to each other. The column may be placed directly on the footing or it may be supported by a pedestal, consisting of a short column that is wider than the supported column. The pedestal helps to reduce the so-called *punching-shear effect* in the footing; it also slightly reduces the edge cantilever distance and, thus, the magnitude of bending in the footing. The pedestal, therefore, allows for a thinner footing and slightly less footing reinforcement. However, another reason for using a pedestal may be to raise the bottom of the supported column above the ground, which is important for columns of wood and steel.

The design of a column footing is based on the following considerations:

Maximum Soil Pressure.   The sum of the unfactored, superimposed load on the footing and the unfactored weight of the footing must not exceed the limit for bearing pressure on the supporting soil material. The required total plan area of the footing is derived on this basis.

Design Soil Pressure.   By itself, simply resting on the soil, the footing does not generate shear or bending stresses. These are developed only by the superimposed load. Thus, the soil pressure to be used for designing the footing is determined by dividing the factored, superimposed load by the actual chosen plan area of the footing.

Control of Settlement.   Where buildings rest on highly compressible soil, it may be necessary to select footing areas that ensure a uniform settlement of all the building foundation supports. For some soils, long-term settlement under dead load only may be more critical in this regard and must be considered, along with maximum soil pressure limits.

Size of the Column.   The larger the column, the less will be the shear and bending stresses in the footing because these are developed by the cantilever effect of the footing projection beyond the edges of the column.

Shear Capacity Limit for the Concrete.   For square-plan footings, this is usually the only critical stress in the concrete. To achieve an economical design, the footing thickness is usually chosen to reduce the need for reinforcement.

Although small in volume, the steel reinforcement is a major cost factor in reinforced concrete construction. This generally rules against any concerns for flexural compression stress in the concrete. As with wall footings, the factored load is used when determining footing thickness and any required reinforcement.

Flexural Tension Stress and Development Length for the Bars.
These are the main concerns for the steel bars, on the basis of the cantilever bending action. It is also desired to control the spacing of the bars between some limits.

Footing Thickness for Development of Column Bars.   When a footing supports a reinforced concrete or masonry column, the compressive force in the column bars must be transferred to the footing by development action (called *doweling*), as discussed in Chapter 8. The thickness of the footing must be adequate for this purpose.

The following example illustrates the design process for a simple, square column footing:

Example 2.   Design a square column footing for the following data:

- Column load = 200 kips [890 kN] dead load and 300 kips [1334 kN] live load
- Column size = 15 in. [380 mm] square
- Maximum allowable soil pressure = 4000 psf [191 kPa]
- Concrete design strength = 3000 psi [20.7 MPa]
- Yield stress of steel reinforcement = 40 ksi [276 MPa]

Solution: A quick guess for the footing size is to divide the load by the maximum allowable soil pressure. Thus,

$$A = \frac{500}{4} = 125 \text{ ft}^2, \qquad w = \sqrt{125} = 11.2 \text{ ft } [3.41 \text{ m}]$$

This does not allow for the footing weight, so the actual size required will be slightly larger. However, it gets the guessing quickly into the approximate range.

For a footing this large, the first guess for the footing thickness is a real shot in the dark. However, any available references to other footings designed for this range of data will provide a reasonable first guess.

Try $h = 31$ in. [787 mm]. Then footing weight $= (31/12)(150) = 388$ psf [18.6 kPa]. Net usable soil pressure $= 4000 - 388 = 3612$ psf [173 kPa]. The required plan area of the footing is, thus,

$$A = \frac{500,000}{3612} = 138.4 \text{ ft}^2 \, [12.9 \text{ m}^2]$$

and the required width for a square footing is

$$w = \sqrt{138.4} = 11.76 \text{ ft} \, [3.58 \text{ m}]$$

Try $w = 11$ ft 9 in., or 11.75 ft. Then design soil pressure $= 500,000/(11.75)^2 = 3622$ psf [173 kPa].

For determining reinforcement and footing thickness, a factored soil pressure is needed.

$$P_u = 1.2 \times P_{DL} + 1.6 \times P_{LL} = 1.2(200) + 1.6(300) = 720 \text{ kips } [3202 \text{ kN}]$$

and

$$W_u = \frac{720}{(11.75)^2} = 5.22 \text{ ksf or } 5220 \text{ psf } [250 \text{ kPa}]$$

Determination of the bending force and moment is as follows (see Figure 13.5):

Bending force:

$$F = 5220 \times \frac{63}{12} \times 11.75 = 322,000 \text{ lb } [1432 \text{ kN}]$$

Bending moment:

$$M_u = 322,000 \times \frac{63}{12} \times \frac{1}{2} = 845,000 \text{ ft-lb } [1146 \text{ kN-m}]$$

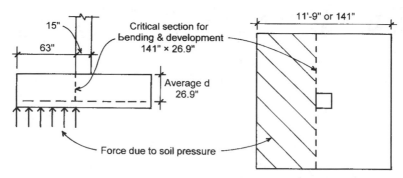

Figure 13.5 Considerations for bending and bar development in the column footing.

and for design

$$M_t = \frac{M_u}{\phi} = \frac{845{,}000}{0.9} = 939{,}000 \text{ ft-lb } [1273 \text{ kN-m}]$$

This bending moment is assumed to operate in both directions on the footing and is provided for with similar reinforcement in each direction. However, it is necessary to place one set of bars on top of the perpendicular set, as shown in Figure 13.6, and there are, thus, different effective depths in each direction. A practical procedure is to use the average of these two depths, that is, a depth equal to the footing thickness minus the 3-in. cover and one bar diameter. This will theoretically result in a minor overstress in one direction, which is compensated for by a minor understress in the other direction.

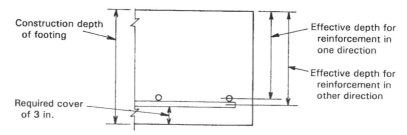

Figure 13.6 Consideration for effective depth of the column footing with two-way reinforcement.

It is also necessary to assume a size for the reinforcing bar in order to determine the effective depth. As with the footing thickness, this must be a guess unless some reference is used for approximation. Assuming a No. 9 bar for this footing, the effective depth, thus, becomes

$$d = h - 3 - (\text{bar } D) = 31 - 3 - 1.13 = 26.87 \text{ in., say } 26.9 \text{ in. } [683 \text{ mm}]$$

The section resisting the bending moment is one that is 141 in. wide and has a depth of 26.9 in. Using a resistance factor for a balanced section from Table 6.1, the balanced moment capacity of this section is determined as follows:

$$M_R = Rbd^2 = \frac{1149 \times 141 \times (26.9)^2}{12} = 9{,}770{,}000 \text{ ft-lb } [13.25 \text{ MN-m}]$$

which is more than ten times the required moment.

From this analysis, it may be seen that the compressive bending stress in the concrete is not critical. Furthermore, the section may be classified as considerably under-reinforced, and a conservative value can be used for j in determining the required reinforcement.

The critical stress condition in the concrete is that of shear, either in beam-type action or in punching action. Referring to Figure 13.7, the investigation for these two conditions is as follows:

For beam-type shear (Figure 13.7a):

$$V_u = 5220 \times 11.75 \times \frac{36.1}{12} = 185{,}000 \text{ lb } [823 \text{ kN}]$$

For the shear capacity of the concrete:

$$V_c = 2\sqrt{f'_c}(b \times d) = 2\sqrt{3000}\,(141 \times 26.9) = 415{,}000 \text{ lb } [1846 \text{ kN}]$$

$$\phi V_c = 0.75(415{,}000) = 311{,}000 \text{ lb } [1383 \text{ kN}]$$

For punching shear (Figure 13.7b):

$$V_u = 5220\left[(11.75)^2 - \left(\frac{41.9}{12}\right)^2\right] = 657{,}000 \text{ lb } [2922 \text{ kN}]$$

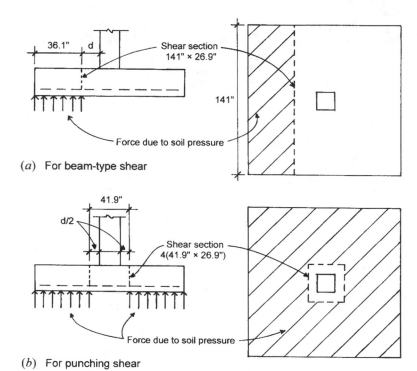

Figure 13.7 Considerations for the two forms of shear development in the column footing: (a) beam-type shear, and (b) punching shear.

Shear capacity of the concrete is

$$V_c = 4\sqrt{f'_c}(b \times d) = 4\sqrt{3000}\,(4 \times 41.9)(26.9) = 988{,}000\ \text{lb}\ [4395\ \text{kN}]$$
$$\phi V_c = 0.75 \times 988{,}000 = 741{,}000\ \text{lb}\ [3296\ \text{kN}]$$

Although the beam shear force is low, the punching shear force is just barely short of the limit, so the 31-in. thickness is indeed the minimum allowable dimension.

Using an assumed value of 0.9 for j, the area of steel required is determined as

$$A_s = \frac{M}{f_y jd} = \frac{939{,}000 \times 12}{40{,}000 \times 0.9 \times 26.9} = 11.64\ \text{in.}^2\ [7510\ \text{mm}^2]$$

TABLE 13.5   Reinforcement Alternatives for the Column Footing

Number and Size of Bars	Area of Steel Provided (Required = 11.64 in.^{2})		Required Development Length^a		Center-to-Center Spacing	
	in.^{2}	mm^2	in.	mm	in.	mm
20 No. 7	12.0	7742	32	813	7.0	178
15 No. 8	11.85	7646	37	940	9.5	241
12 No. 9	12.0	7742	42	1067	12.1	307
10 No. 10	12.7	8194	47	1194	14.7	373
8 No. 11	12.48	8052	52	1321	19.0	483

^{a}From Table 13.8, values for "other bars," $f_y = 40$ ksi, $f'_c = 3$ ksi.

A number of combinations of bar size and number may be selected to satisfy this area requirement. A range of possible choices is shown in Table 13.5. Also displayed in the table are data relating to two other considerations for the bar choice: the center-to-center spacing of the bars and the development lengths required. Spacings given in the table assume the first bar to be centered at 4 in. from the footing edge. Maximum spacing should be limited to 18 in. and minimum to about 6 in.

Required development lengths are taken from Table 8.1. The development length available is a maximum of the distance from the column face to the footing edge minus a 2-in. cover—in this case, a distance of 61 in.

Inspection of Table 13.5 reveals that all the combinations given are acceptable. In most cases, designers prefer to use the largest possible bar in the fewest number because handling of the bars is simplified with fewer bars, which is usually a savings of labor time and cost.

Although the computations have established that the 31-in. dimension is the least possible thickness, it may be more economical to use a thicker footing with less reinforcement, assuming the usual ratio of costs of concrete and steel. In fact, if construction cost is the major determinant, the ideal footing is the one with the lowest combined cost for excavation, forming, concrete, and steel.

One possible limitation for the footing reinforcement is the total percentage of steel. If this is excessively low, the section is hardly being reinforced. The ACI Code stipulates that the minimum reinforcement be the same as that for temperature reinforcement in slabs, a percentage of $0.002A_g$ for Grade 40 bars and $0.0015A_g$ for Grade 60 bars. For this footing cross section of 141 in. by 31 in. with Grade 40 bars, this means an area of

$$A_s = 0.002(141 \times 31) = 8.74 \text{ in.}^2 \text{ [5639 mm}^2\text{]}$$

A number of other considerations may affect the selection of footing dimensions, such as the following:

Restricted Thickness. Footing thickness may be restricted by excavation problems, water conditions, or the presence of undesirable soil materials at lower strata. Thickness may be reduced by the use of pedestals, as discussed in Section 13.8.

Need for Dowels. When the footing supports a reinforced concrete or masonry column, dowels must be provided for the vertical column reinforcement, with sufficient extension into the footing for development of the bars. This problem is discussed in Section 8.5.

Restricted Footing Width. Proximity of other construction or close spacing of columns sometimes makes it impossible to use the required square footing. For a single column, a possible solution is the use of an oblong (called a *rectangular*) footing. For multiple columns, a combined footing is sometimes used. A special footing is the cantilever footing, used when footings cannot extend beyond the building face. An extreme case occurs when the entire building footprint must be used in a single large footing, called a *mat foundation*. See Section 13.7.

Table 13.6 yields the allowable superimposed load for a range of predesigned footings and soil pressures. This material has been adapted from more extensive data in *Simplified Design of Building Foundations* (Reference 7). Designs are given for footings using concrete strength of 3000 psi. Figure 13.8 indicates the symbols used for dimensions in Table 13.6. As discussed for the wall footings and elsewhere in this book, a low design strength of 2000 psi may sometimes be used to avoid the necessity for the usual code-required field testing of concrete. However, no structural concrete should be specified with a strength less than 3000 psi.

Problem 13.6.A.   Design a square footing for a 14-in. [356 mm] square column and a superimposed dead load of 100 kips [445 kN] and a live load of 100 kips [445 kN]. The maximum permissible soil pressure is 3000 psf [144 kPa]. Use concrete with a design strength of 3 ksi [20.7 MPa] and Grade 40 reinforcing bars with a yield strength of 40 ksi [276 MPa].

TABLE 13.6   Safe Loads for Square Column Footings[a] (see Figure 13.8)

Maximum Soil Pressure (psf)	Minimum Column Width, t (in.)	Service Load on Footing (kips)	Dimensions		Reinforcement Each Way
			h (in.)	w (ft)	
1000	8	7	10	3	3 No. 2
	8	10	10	3.5	3 No. 3
	8	14	10	4	4 No. 3
	8	17	10	4.5	4 No. 4
	8	21	10	5	4 No. 5
	8	31	10	6	4 No. 6
	8	42	11	7	6 No. 6
1500	8	12	10	3	3 No. 3
	8	16	10	3.5	3 No. 4
	8	22	10	4	4 No. 4
	8	27	10	4.5	4 No. 5
	8	34	10	5	5 No. 5
	8	49	12	6	5 No. 6
	8	65	13	7	5 No. 7
	8	84	15	8	7 No. 7
	8	105	17	9	8 No. 7
2000	8	16	10	3	3 No. 3
	8	23	10	3.3	3 No. 4
	8	30	10	4	5 No. 4
	8	38	10	4.5	5 No. 5
	8	46	11	5	4 No. 6
	8	66	13	6	6 No. 6
	8	89	15	7	6 No. 7
	8	114	17	8	8 No. 7
	8	143	19	9	7 No. 8
	10	175	20	10	9 No. 8
3000	8	25	10	3	3 No. 4
	8	35	10	3.5	3 No. 5
	8	45	11	4	4 No. 5
	8	57	12	4.5	4 No. 6
	8	71	13	5	5 No. 6
	8	101	15	6	7 No. 6
	10	136	17	7	7 No. 7
	10	177	20	8	7 No. 8
	12	222	21	9	9 No. 8
	12	272	24	10	9 No. 9
	12	324	26	11	10 No. 9
	14	383	28	12	10 No. 10

(*continued*)

TABLE 13.6 *(Continued)*

Maximum Soil Pressure (psf)	Minimum Column Width, t (in.)	Service Load on Footing (kips)	Dimensions h (in.)	w (ft)	Reinforcement Each Way
4000	8	34	10	3	4 No. 4
	8	47	11	3.5	4 No. 5
	8	61	12	4	5 No. 5
	8	77	13	4.5	5 No. 6
	8	95	15	5	5 No. 6
	8	136	18	6	6 No. 7
	10	184	20	7	8 No. 7
	10	238	23	8	8 No. 8
	12	300	25	9	8 No. 9
	12	367	27	10	10 No. 9
	14	441	29	11	10 No. 10
	14	522	32	12	11 No. 10
	16	608	34	13	13 No. 10
	16	698	37	14	13 No. 11
	18	796	39	15	14 No. 11

*a*Service loads do not include the weight of the footing, which has been deducted from the total bearing capacity. Service load is considered 40% dead load and 60% live load. Grade 40 reinforcement. $f'_c = 3$ ksi.

Problem 13.6.B. Same as Problem 13.6.A, except column is 18 in. [457 mm], dead load is 200 kips [890 kN], and live load is 300 kips [1334 kN]. Permissible soil pressure is 4000 psf [192 MPa].

13.7 SPECIAL COLUMN FOOTINGS

Situations commonly occur in which a simple, square footing may not be indicated for a column. The following are some special forms for column footings for frequently encountered situations:

Rectangular Footing. When soil design pressures are low or a column must be placed close to some other construction, it may be necessary to use a footing that is oblong in plan, rather than square, a form referred to as a *rectangular footing*. Design is performed essentially as for a square footing, except that there are special requirements for placing of the reinforcement in the short direction.

Combined Footing. When two or more columns are placed close together in the building plan, a single footing called a *combined footing* is

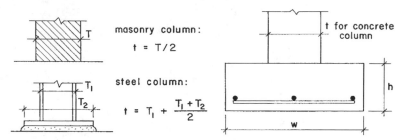

masonry column:

$t = T/2$

steel column:

$t = T_1 + \dfrac{T_1 + T_2}{2}$

t for concrete column

Figure 13.8   Reference figure for Table 13.6.

sometimes used. An example of such a footing for two equally loaded columns is shown in Figure 13.9. In this case, a simple oblong footing is placed symmetrically beneath the columns and is designed as a double cantilevered beam. If the columns are not equally loaded, the footing may be shifted to have its plan centroid coincide with that of the column loads, or some other form, such as those shown in Figure 13.10, may be used.

Cantilever Footing.   A common situation that occurs with buildings on tight urban sites is that the edge of the building is placed very close to the property line. If a column occurs at the edge of the building, a conventional footing would likely extend a considerable distance beyond the edge of the building and over the property line. In such a case, one solution is to use a *cantilever footing,* also called a *strap footing,* consisting of a combined footing supporting the exterior column and an adjacent interior column. Such a footing is shown in Figure 13.11, with a stiffening stem wall to form a T-beam action for the major bending that occurs midway between the columns.

13.8   PEDESTALS

A *pedestal* (also called a *pier*) is defined by the ACI Code as a short compression member whose height does not exceed three times its width. Pedestals are frequently used as transitional elements between columns and the bearing footings that support them. Figure 13.12 shows the use of pedestals with both steel and reinforced concrete columns.

The most common reasons for use of pedestals are:

1. To spread the load on top of the footing. This may relieve the intensity of direct bearing pressure on the footing or may simply per-

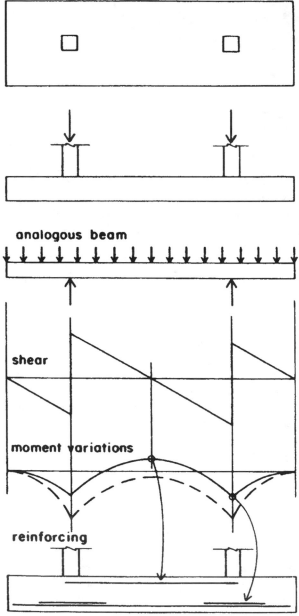

Figure 13.9 Actions of a symmetrically loaded combined footing for two columns.

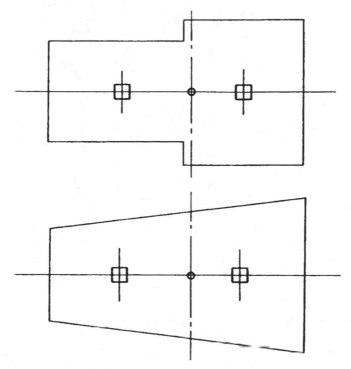

Figure 13.10 Plan variations for a combined footing for two columns with different loads.

mit a thinner footing with less reinforcement due to the wider column.

2. To permit the column to terminate at a higher elevation, where footings must be placed at depths considerably below the lowest parts of the building. This is generally most significant for steel columns.

3. To provide for the required development length of reinforcing in reinforced concrete columns, where footing thickness is not adequate for development within the footing.

4. To effect a transition between a column with very high concrete strength and a footing with only moderate concrete strength.

Figure 13.12*d* illustrates the third situation described. Referring to Table 8.3, we may observe that a considerable development length is re-

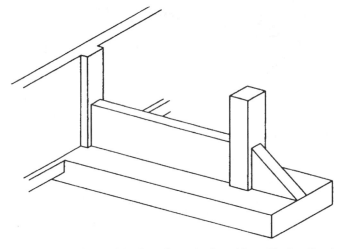

Figure 13.11   Use of a combined cantilever footing with a stiffening rib, where the footing cannot project beyond the outside face of the basement wall.

quired for large-diameter bars made from high grades of steel. If the minimum required footing does not have a thickness that permits this development, a pedestal may offer a reasonable solution. However, there are many other considerations to be made in the decision, and the column-reinforcing problem is not the only factor in this situation.

If a pedestal is quite short with respect to its width (see Figure 13.12*e*), it may function in essentially the same way as a column footing, with significant values for shear and bending stresses. This condition is likely to occur if the pedestal width exceeds twice the column width, and the pedestal height is less than one-half of the pedestal width. In such cases, the pedestal must be designed by the same procedures used for an ordinary column footing.

The following example illustrates the procedure for the design of a pedestal for a reinforced concrete column:

Example 3.   A 16-in. [406 mm] square tied column with f'_c of 4 ksi [27.6 MPa] is reinforced with No. 10 bars of Grade 60 steel ($f_y = 60$ ksi [414 MPa]). The column axial load is 100 kips [445 kN] dead load and 100 kips [445 kN] live load, and the allowable maximum soil pressure is 4000 psf [192 kPa]. Design a pedestal, using $f'_c = 3$ ksi [20.7 MPa] and Grade 40 reinforcement with $f_y = 40$ ksi [276 MPa].

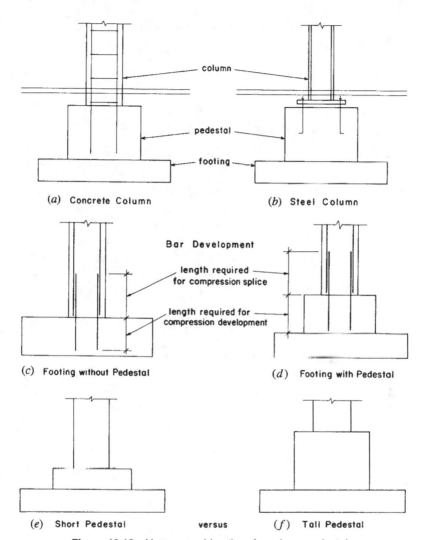

(a) Concrete Column   (b) Steel Column

Bar Development

length required
for compression splice

length required for
compression development

(c) Footing without Pedestal   (d) Footing with Pedestal

(e) Short Pedestal   versus   (f) Tall Pedestal

Figure 13.12 Usage considerations for column pedestals.

Solution: For an approximate idea of the required footing, we may refer to Table 13.6 and observe the following:

- 8-ft [2.44 m] square footing, 23 in. [584 mm] thick, nine No. 8 each way
- Allowable load on footing: 238 kips [1059 kN]
- Designed for column width of 10 in. [254 mm]

From Table 8.3, for No. 10 bar, Grade 60, we observe the following:

$$f'_c = 3 \text{ ksi [20.7 MPa]}, \ l_d = 28 \text{ in. [711 mm]}$$

From these observations, we may conclude that:

1. The minimum required footing for the 16-in. column with a total service load of 200 kips will be slightly smaller than that taken from the table. Thus, it will not be adequate for development of the column bars.
2. If a pedestal is used, it must be at least 28 in. high to develop the column bars.
3. With a pedestal slightly wider than the column, the footing thickness may be additionally reduced if shear stress is the critical design factor for the footing thickness.

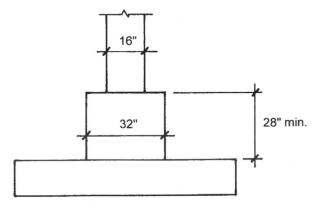

Figure 13.13 Pedestal example.

One option in this case is to simply forget about a pedestal and increase the footing thickness to that required for development of the column bars. This means an increase from around 20 in. up to 31 in., giving the necessary 28 in. of development plus 3 in. of cover.

If it is desired to use a pedestal, consider the use of the one shown in Figure 13.13. The 28-in. height shown is the minimum established previously for the development of the column bars. The height could be increased to as much as 96 in. (three times the width), if it is desired for other reasons. One such reason may be the presence of a better soil for bearing at a lower elevation.

A potential concern is the direct bearing of the column on the pedestal. If the pedestal is designed as an unreinforced member, the ACI Code permits a maximum bearing strength of

$$\phi B_u = 0.55(0.85 f'_c A_1)$$

which may be increased by a factor of $\sqrt{A_2/A_1}$.

A_1 is the actual bearing area (in our case the 16-in. square column area), and A_2 is the area of the pedestal cross section. The maximum usable value for $\sqrt{A_2/A_1}$ is 2.

For the pedestal, there are two concerns: the bearing of the column on top of the pedestal and the column compression capacity of the pedestal. For the factored load, we determine

$$P_u = 1.2(DL) + 1.6(LL) = 1.2(100) + 1.6(100) = 280 \text{ kips } [1245 \text{ kN}]$$

The bearing capacity is

$$\phi B_n = 0.55(0.85 f'_c A_1)\sqrt{A_2/A_1}$$

$$= 0.55[0.85 \times 3 \times (16)^2]\sqrt{(32)^2/(16)^2}$$

$$= 718 \text{ kips } [3194 \text{ kN}]$$

which is considerably larger than required.

For the capacity of the pedestal as an unreinforced column

$$\phi P_n = 0.55(0.6 f'_c A_2)\left[1 - \left(\frac{L_c}{32h}\right)^2\right]$$

$$= 0.55(0.6 \times 3 \times (32)^2) \left[1 - \left(\frac{28}{32 \times 32} \right)^2 \right]$$

$$= 1013 \text{ kips } [4506 \text{ kN}]$$

which is also greater than required. In this equation, L_c is the unbraced height of the pedestal, and h is the pedestal thickness (width).

The only times that these conditions are likely to be critical is when a pedestal with very low f'_c supports a column with very high f'_c and the pedestal width is only slightly greater than the column width. When the pedestal supports a steel column, however, this condition may be the basis for establishing the required width of the pedestal.

Another consideration for bearing is that of the pedestal on the footing. This is also not critical for our example.

If the pedestal height exceeds its width, a minimum column reinforcing of not less than $A_s = 0.005 A_g$ is recommended. This should be installed with at least four bars, one in each corner, and a set of loop ties, just as with an ordinary tied column. For our short pedestal, this is of questionable necessity.

With the wide pedestal, the footing thickness can be reduced considerably if the minimum thickness for shear strength is desired. The cost of the thinner footing with less reinforcement, plus the cost of the pedestal, must be compared with the cost of a typical square footing of greater thickness. This typically is close to a trade-off, so other reasons for having a pedestal usually govern its use.

Design for the footing with the pedestal is essentially the same as the typical design for a column footing, as described in Section 13.6.

Problem 13.8.A.   An 18-in. [457 mm] square tied column with $f'_c = 4$ ksi [27.6 MPa] is reinforced with No. 11 bars of Grade 60 steel with $f_y = 60$ ksi [414 MPa]. The column axial load is 120 kips [534 kN] dead load and 140 kips [623 kN] live load. Using $f'_c = 3$ ksi [20.7 MPa] and Grade 40 bars with $f_y = 40$ ksi [276 MPa], design a pedestal.

Problem 13.8.B.   Same as Problem 13.8.A, except column is 22 in. [559 mm] square, reinforcement is No. 14 bars, dead load is 300 kips [1334 kN], and live load is 190 kips [845 kN].

13.9 FOUNDATION WALLS AND GRADE BEAMS

Foundation walls extend below the ground surface and effect a transition between the aboveground building construction and the buried elements of the foundation system. As with other ground-contacting construction, they usually consist of concrete or masonry. The architectural and structural functions of foundation walls vary considerably, depending on the type of foundation, the size of the building and its form of construction, local climate and soil conditions, and whether or not they serve to form a basement or other belowground space.

Figure 13.14 shows some typical situations for the use of foundation walls.

With no basement, the foundation walls are not actually walls in the usual sense, but rather simply serve to keep the aboveground construction

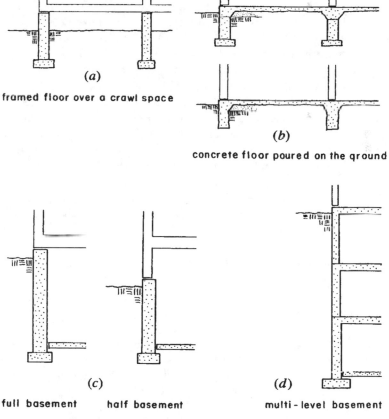

(*a*)

framed floor over a crawl space

(*b*)

concrete floor poured on the ground

(*c*)

full basement   half basement

(*d*)

multi-level basement

Figure 13.14 Typical situations for basement walls.

truly above ground, and effect the transition to the footings. The avoidance of ground contact is especially important for building construction of wood or steel. As shown in Figure 13.14*b,* a principal difference has to do with the actual depth of the footings below the exterior ground surface (commonly called the *grade*). Where a considerable depth of frost occurs, or where good soil is at some distance below grade, the form of construction shown in the upper part of Figure 13.14*b* may be used, with a short foundation wall formed and separately cast. However, for light vertical loads (one-story buildings, mostly) and no frost problems, the wall and footing are sometimes combined into a single element, called a *grade beam,* as shown in the lower part of Figure 13.14*b.* If the grade beam and floor slab can be continuously cast in a single concrete pour, both time and money are saved.

Where no basement occurs, there are two possibilities for the construction. The first, as shown in Figure 13.14*a,* consists of extending foundation walls or individual piers from the top of the footings to some distance above grade. This results in the creation of a space—called a *crawl space*—between the ground surface and the underside of the supported floor construction above. The other alternative is to use a floor consisting of a concrete paving slab; this system is described in Section 14.2.

When a basement is required, foundation walls are usually quite tall. An exception is the case of a half-basement, occurring when the basement floor is only a short distance below exterior grade (see Figure 13.14*c*). The ordinary basement wall is discussed in Section 12.3. When multiple levels occur below grade, the foundation walls develop a multistory structure, as shown in Figure 13.14*d.*

In addition to their usual function of providing an aboveground support for the building construction, foundation walls often serve a variety of functions, such as the following (see Figure 13.15):

Load Distribution or Equalizing. Walls of some length typically serve as rather stiff beams, distributing loads to their continuous footings. They also help to compensate for uneven soil pressures on the relatively thin footings, which is a common occurrence.

Beam Actions. Walls may be used as spanning members, carrying both their own weight and some supported loads. This action may occur when bearing soils vary considerably along the wall length, as shown in Figure 13.15*b.* It is also typically the case with foundations using deep elements (piles or piers) that occur at spaced intervals along the walls. It may also be the case for bearing foundations with building columns car-

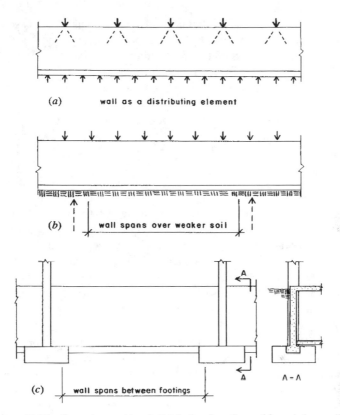

(a)   wall as a distributing element

(b)   wall spans over weaker soil

(c)   wall spans between footings

Figure 13.15 Spanning and load-distributing functions of foundation walls.

rying considerable vertical load. In the latter case, as shown in Figure 13.15c, column footings will be large, and the walls will tend to span between the column footings, rather than bear on their own narrow and shallower wall footings.

Wall and Column Interaction.   When building columns occur in the same plane as foundation walls, many different relationships are possible. If the columns and walls are continuously cast, some load sharing is unavoidable.

Transfer of Lateral Loads.   Regardless of the type of lateral-load-resisting system for the building, the lateral forces must eventually be transferred to the ground. Horizontal forces, as well as some possible up-lift forces, may be transferred first to supporting foundation walls. The complete system of foundation walls may also serve as a lateral-load-distributing system for isolated foundations.

In one way or another, most foundation walls serve some of these structural functions. The dead weight of the foundation construction is often the basic anchorage resistance for light buildings subjected to high winds. The design of a basement wall for horizontal soil pressure is illustrated in Section 12.3.

Grade Beams

A grade beam is a linear foundation element used to span across bearing supports at grade level. The supports may consist of isolated footings, pile groups, or piers, so the grade beam truly acts as a spanning beam. However, another frequent application of the grade beam is to form a distributing element for uneven soil supporting conditions. For the latter purpose, most foundation walls serve some grade beam function by distributing the loads they support evenly to the soil, as well as spanning across any uneven spots in the soil.

Indeed, most grade beams consist of the basement walls or grade walls used with wall footings for the building base. For this reason, it is common practice to place some continuous steel reinforcement in the tops and bottoms of all foundation walls, to permit them to function as continuous, reinforced concrete beams for bending moments of either positive or negative form.

A special grade beam is one that is sometimes used for slab floors on grade where frost actions are not critical. As shown in Figure 13.16a, this element may be cast continuously with the floor slab. It functions as the wall footing, a grade beam for the usual purposes, and a stiffener for the edge of the slab. Where the construction sequence requires the walls to be built before the floor slab is cast, it is possible to use the detail shown in Figure 13.16b, which permits the same structural functioning as the construction shown in Figure 13.16a.

13.10 DEEP FOUNDATIONS

Far and away the cheapest, simplest, most common foundation is the shallow bearing footing. However, the use of bearing pressure on yielding soil has some limitations that must be recognized; these include the following:

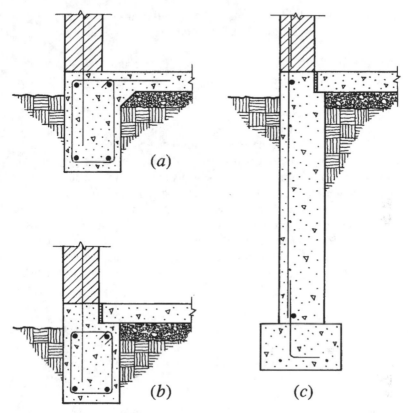

Figure 13.16 Forms of grade beams: (a) as a thickened edge of a slab, serving as a footing for a wall, (b) as a variation of (a) permitting wall construction to precede floor slab construction, and (c) as a foundation wall reinforced to act as a spanning member.

Soil Strength.   Load capacity of soils—even the strongest ones—is limited. Good structural wood, for example, typically has a usable compression strength of around 1000 lb/sq in., which is weak in comparison to steel or superstrength concrete. But a soil is considered superstrength if it can sustain as much as 10,000 lb/sq ft, or about 70 lb/sq in. Excessive loading, from heavy construction, tall buildings, or long span structures, can require footings with massive plan areas to keep soil pressure low.

Settlement.   Soil is compressible; the greater the pressure, the more the deformation, which accumulates as downward vertical movement (settlement). With large deposits of clays, settlements may be progressive over time, accumulating in feet, not inches.

Instability.   Soils with discrete particles not generally tightly bonded or cemented are subject to erosion, consolidation, organic or general chemical decomposition, and massive shrinkage or expansion due to major changes in water content.

For these or other reasons, use of deep foundations may be favored. Actually, the simple reason for choosing what is almost always a more expensive foundation system is usually that the bearing capacity and assurance of limited settlement needed are not available at the location where bearing footings would need to be located.

Many basic forms of deep foundations are possible, and many special ones are available as proprietary systems. The four types commonly used for building foundations are those shown in Figure 13.17. Individual systems become popular regionally due to a combination of availability, appropriateness to local problems, and marketing concentration by companies.

Piles

Piles are elements driven into the ground in the manner of large nails. Various special means of advancing piles are possible, including vibration, but the usual means is simply to pound them in. The two basic types are:

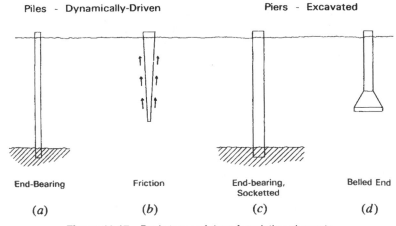

Figure 13.17   Basic types of deep foundation elements.

Friction Piles.   These simply develop resistance to being farther pushed into the soil due to friction on their surfaces—directly analogous to a nail in wood. Their capacity is inferred from the difficulty encountered in driving them the last few feet.

End-Bearing Piles.   These are driven until their ends encounter extreme resistance—usually because of rock, but sometimes simply because of a very hard layer of soil beneath much softer ones.

Piles take various forms and use various combinations of wood, steel, and concrete. Various proprietary systems consist of the driving of a steel shell, using a solid steel core during the driving (called a *mandrel*). The mandrel is withdrawn when resistance is adequate, and the hollow shell is filled with concrete. In some soils, the shell can also be withdrawn during the placing of the concrete. In other systems the shell itself is driven, usually for development of an end-bearing condition.

The oldest form of pile, and still one of the cheapest, is the simple timber pole. Rot or consumption by living organisms is a problem that must be dealt with, but these are still widely used in regions where cheap timber poles are readily available. For permanent, major construction, however, steel and concrete are now favored—sometimes simply because of their larger load capacities.

Pile Caps.   Pile caps function much like column footings and are usually close in size to column footings for the same supported load. As shown in Figure 13.18, they consist essentially of footings built on top of a cluster of piles. Pile layouts typically follow classical patterns, based on the number of piles in the group. Typical plan layouts are shown in Figure 13.19. Special layouts may be used for groups that carry walls, elevator towers, large machinery, freestanding towers, and other structures.

The three-pile group is ordinarily preferred as the minimum for a single column due to the difficulty of accurately locating the tops of driven piles. However, lateral bracing with foundation walls or struts may permit use of a two-pile cluster.

Excavated Shafts, Piers, or Caissons

These are literally tall columns of concrete, constructed in an excavated, shaftlike hole. They function in a way similar to end-bearing piles, with ends inserted into rock, called *socketed ends* (see Figure 13.17*c*), or

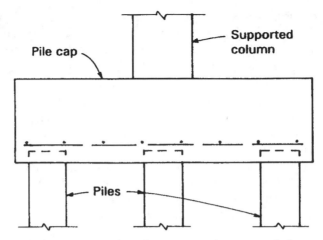

Figure 13.18   Reinforced concrete cap for a group of piles.

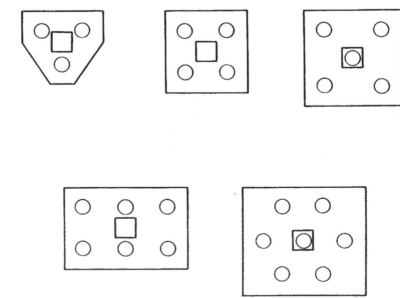

Figure 13.19   Typical pile layouts and cap plan forms.

widened to bear on hard soil, called *belled ends* (see Figure 13.17*d*). They may range in size from small (12 in. diameter) up to gigantic—the latter for highrise buildings, large bridge piers, and so on.

Where soil conditions permit, piers of small-to-moderate size and relatively short length may be excavated by drilling, in the general manner used for postholes or water wells. Large piers, however, must simply be dug out, with lining of the shaft walls installed as the hole is advanced.

The term *caisson*, which is still commonly used to describe this element, actually comes from a method that was developed many years ago for advancing excavations for large piers in soft soils. This consists of building a working chamber (the *caisson*, French for "box") with no bottom, and then digging out the soil to steadily lower the chamber. Once lowered to its desired location, the chamber becomes the bottom of the pier. This method is still used, but mostly for large bridge piers under water.

Choice of foundation systems involves many considerations, including factors relating to soil conditions, excavation requirements, size and type of the building project, local availability of foundation work, and experiences with construction.

Construction of Deep Foundations

The general planning of buildings that are supported on piles or piers, rather than simple bearing footings, requires some special considerations. These may differ for various particular forms of the deep foundation elements, but they have some typical conditions, including the following:

Point Supports. Piers or clusters of piles must be spaced some distance apart. This does not generally affect planning of columns, but does cause differences for walls and slabs on grade.

Minimum Load Unit. Footings can be made quite small, appropriate to their required load-carrying tasks. Piles and piers typically have a minimum size, which may represent a significant excess load capacity. This factor is even more critical for column loads on piles because the minimum pile cluster usually consists of two or three piles and it is even more critical for one-story buildings where the only column loads are from light roofs.

Use of Heavy Construction Equipment.   Installation of piles and drilled piers requires heavy equipment, which cannot economically be transported great distances for small projects. Access to difficult sites (hillsides, swamps, etc.) may present a problem. In addition, the operation of pile drivers may upset the neighbors.

For further discussion of the design of deep foundations, readers are referred to *Simplified Design of Building Foundations* (Reference 7).

14

MISCELLANEOUS CONCRETE STRUCTURES

Concrete is a versatile material, usable for everything from birdbaths to monumental structures. This chapter presents material for a few of the special uses of reinforced concrete.

14.1 PAVING SLABS

Sidewalks, driveways, and basement floors are typically produced by depositing a relatively thin coating of concrete directly on the ground surface. Although the basic construction process is simple, a number of factors must be considered in developing details and specifications for a paving slab.

Thickness of the Slab

Pavings vary in thickness from a few inches (for residential basement floors) to several feet (for airport landing strips). Although more strength

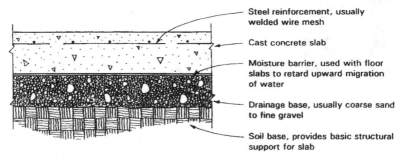

Figure 14.1 Typical concrete paving slab.

is implied by a thicker slab, thickness alone does not guarantee a strong pavement. Also of concern are the reinforcement provided and the character of the developed subbase on which the concrete is poured (see Figure 14.1). The minimum slab thickness commonly used in building floor slabs is $3\frac{1}{2}$ in. This relates specifically to the actual dimension of a nominal wood two-by-four and simplifies forming the edges of a slab pour. Following the same logic, the next-size jump would be to a $5\frac{1}{2}$-in. thickness, which is the dimension of a nominal two-by-six.

The $3\frac{1}{2}$-in.-thick slab is usually considered adequate for interior floors not subjected to wheel loadings or other heavy structural demand. At this thickness, usually provided with very minimal reinforcing, the slab has relatively low resistance to bending and shear effects of concentrated loads. Thus, bearing walls, columns, and heavy items of equipment should be provided with separate footings.

The $5\frac{1}{2}$-in.-thick slab is adequate for heavier distributed live loads. For other situations involving very heavy loads—especially concentrated ones—thicker pavements should be used, although thickness alone is not sufficient, as mentioned previously.

Reinforcement

Thin slabs are ordinarily reinforced with welded wire mesh. The most commonly used meshes are those with a square pattern of wires—typically 4- or 6-in. spacings—with the same wire size in both directions. This reinforcing is generally considered to provide only for shrinkage and temperature effects, and to add little to the flexural strength of the slab. The minimum mesh, commonly used with the $3\frac{1}{2}$-in. slab, is a 6×6

10/10, which denotes a mesh with No. 10 wires at 6 in. on center in each direction. For thicker slabs, the wire gage should be increased or two layers of mesh should be used.

Small-diameter reinforcing bars are also used for slab reinforcement, especially with thicker slabs. These are generally spaced at greater distances than the mesh wires and must be supported during the pouring operation. Unless the slab is actually designed to span, this reinforcement is still considered to function primarily for shrinkage and temperature stress resistance. However, because cracking in the exposed top surface of the slab is usually the most objectionable, specifications usually require the reinforcing to be kept some minimum distance from the top of the slab.

Subbase

The ideal subbase for floor slabs is a well-graded soil, ranging from fine gravel to coarse sand with a minimum of fine materials. This material can usually be compacted to a reasonable density to provide a good structural support, while retaining good drainage properties to avoid moisture concentrations beneath the slab. Where ground water conditions are not critical, this base is usually simply wetted down before pouring the concrete, and the concrete is deposited directly on the subbase. The wetting serves somewhat to consolidate the subbase and to reduce the bleeding out of the water and cement from the bottom of the concrete mass.

To reduce further the bleeding-out effect, or where moisture penetration is more critical, a lining membrane is often used between the slab and the subgrade or subbase.

Joints

Building floor slabs are usually poured in relatively small units, in terms of the horizontal dimension of the slab. The main reason is to control shrinkage effects. A full break in the slab, formed as a joint between successive pours, provides for the incremental accumulation of the shrinkage effects. Where larger pours are possible or more desirable, control joints are used. These consist of tooled or sawed joints that penetrate some distance down from the finished top surface.

Surface Treatment

Where the slab surface is to serve as the actual wearing surface, the concrete is usually formed to a highly smooth surface by troweling. This surface may then be treated in a number of ways, such as by brooming it to make it less slippery, or by applying a hardening compound to further toughen the wearing surface. When a separate material—such as tile or a separate concrete fill—is to be applied as the wearing surface, the surface is usually kept deliberately rough. This may be achieved by simply reducing the degree of finished troweling.

Weather Exposure

Once the building is enclosed, interior floor slabs are not ordinarily exposed to exterior weather conditions. In cold climates, however, freezing and extreme temperature ranges should be considered if slabs are exposed to the weather. This may indicate the need for more temperature reinforcing, less distance between control joints, or the use of materials added to the concrete mix to enhance resistance to freezing.

14.2 FRAMED FLOORS ON GRADE

It is sometimes necessary to provide a concrete floor poured directly on the ground in a situation that precludes the use of a simple paving slab and requires the floor structure to have a real structural spanning capability. One of these situations is where a deep foundation is provided for support of walls and columns, and the potential settlement of upper ground masses may result in a breaking up and subsidence of the paving. Another such situation occurs when considerable fill must be placed beneath the floor, and it is not feasible to produce a compaction of this amount of fill to ensure steady support for the floor.

Figure 14.2 illustrates two techniques that may be used to provide what amounts to a framed concrete slab and beam system poured directly on the ground. Where spans are modest and beam sizes not excessive, it may be possible to provide the system in a single pour by simply trenching for the beam forms, as shown in the upper illustration in Figure 14.2. When larger beams are required, the stems of the beams are formed and poured and the slab is poured separately on the fill placed between the beam stems. These two techniques can be blended, of course, with smaller beams trenched in the fill between the large formed beams.

If the system with separately poured beams and slabs is used, it is necessary to provide for the development of shear between the slab and the top of the beam stems. Depending on the actual magnitude of the shear stresses involved, this may be done by various means. If stress is low, it may be sufficient to require a roughening of the surface of the top of the

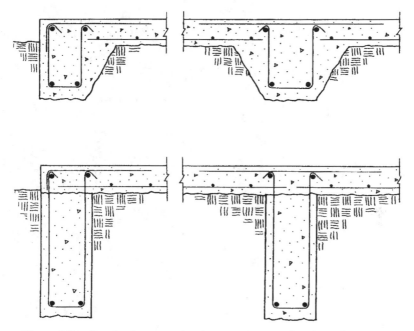

Figure 14.2 Details of concrete framing systems cast directly on the ground.

beam stems. If stress is of significant magnitude, shear keys, similar to those used for shear walls, may be used. If stirrups or ties are used in the beam stems, these will extend across the joint and assist in the development of shear.

14.3 CANTILEVER RETAINING WALLS

Strictly speaking, any wall that sustains significant lateral (horizontal) soil pressure is a *retaining wall*. However, the term is usually used for *cantilever retaining walls,* which are freestanding walls without lateral support at the top. For such a wall, the major design consideration is for the dimension of ground level difference on the two sides of the wall. The range of this dimension establishes the following categories for the retaining structure:

Curbs. These are the shortest freestanding retaining structures. The two most common forms are as shown in Figures 14.3*a* and *b*. The selection between these two forms is made on the basis of whether or not a gutter

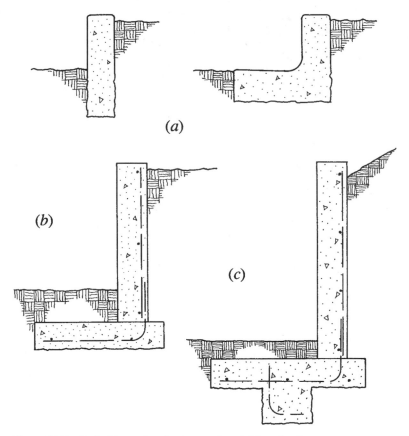

Figure 14.3   Typical forms for short concrete retaining structures

is formed on the low side of the curb. Use of these structures is typically limited to grade-level changes of 2 ft or less.

Short Retaining Walls.   Walls up to about 10 ft in height are usually built as shown in Figure 14.3*b*. These consist of a concrete or masonry wall of uniform thickness. The wall thickness, footing width and thickness, vertical wall reinforcement, and transverse footing reinforcement are all designed for the lateral-load-induced cantilever moments plus the weight of the wall, footing, and earth fill. When the bottom of the footing is a short distance below grade on the low side of the wall and/or the lateral resistance of the soil is low, it may be necessary to use a shear key, as shown in Figure 14.3c.

Tall Retaining Walls.   As the wall height increases, it becomes less feasible to use the simple construction shown in Figure 14.3. The overturning moment increases sharply with increase in the height of the wall. For very tall walls, one modification used is to taper the wall thickness. This allows for the development of a strong cross section for the high bending moment at the bottom of the wall, without an excessive increase of concrete volume. However, as the wall gets very tall, it is often necessary to brace the wall. Bracing may be created with construction, as shown in Figure 14.4, or with tiebacks anchored in the earth behind the wall.

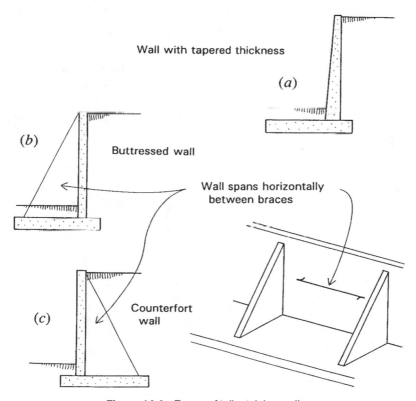

Figure 14.4   Forms of tall retaining walls.

The design of tall retaining walls is beyond the scope of this book. They should be designed with a more rigorous and exact analysis of active and passive soil pressures, and of development of lateral force on the wall. Under ordinary circumstances, however, it is reasonable to design short walls by the equivalent fluid pressure method, which is demonstrated in the following example:

Example.   A short retaining wall is proposed with the profile shown in Figure 14.5. Investigate for the adequacy of the wall dimensions, and select reinforcement for the wall and its footing. Use the following data:

*   Active horizontal soil pressure is 30 psf [1.44 kPa].
*   Soil weight is 100 lb/ft³ [1602 kg/m³].
*   Maximum allowable vertical soil pressure is 1500 psf [71.8 kPa].
*   Concrete strength is 3000 psi [20.7 MPa].
*   Reinforcement is Grade 40 with f_y = 40 ksi [276 MPa].

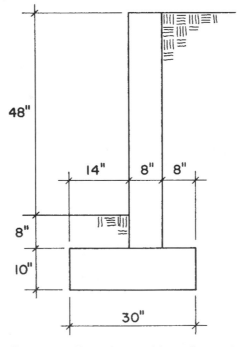

Figure 14.5   Form of the retaining wall example.

Solution: The loading condition used to investigate the vertical soil pressure and the bending in the footing is shown in Figure 14.6. In addition to not exceeding the limit for soil pressure, it is usually desirable to avoid tension stress on the footing-to-earth contact face. For a rectangular footing plan, this means an eccentricity of the load resultant of not more than $\frac{1}{6}$ of the footing width.

Table 14.1 contains the data for determination of the location of the resultant force at the footing bottom. The location of this resultant is found by dividing the sum of the moments about the center of the footing by the sum of the vertical forces, thus

$$e = \frac{5793}{1167} = 4.96 \text{ in. } [126 \text{ mm}]$$

This is compared to $30/6 = 5$ in., so the resultant is just inside of the limiting dimension. Combined soil pressures due to the vertical loads and overturning moment are determined as follows:

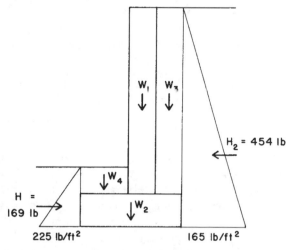

Figure 14.6 Loading for investigation of the footing and soil pressure for the retaining wall example.

TABLE 14.1 Determination of the Eccentricity of the Resultant Force

	Force (lb)	Moment Arm (in.)	Moment (lb-in.)
H_2	454	22	+9988
w_1	466	3	−1398
w_2	312	0	0
w_3	311	11	−3421
w_4	78	8	+624
Totals:	1167		+5793

$$p = \frac{N}{A} \pm \frac{M}{S}$$

where:

- $N =$ the total vertical force
- $A =$ the plan area of the footing
- $M =$ the net moment about the footing center
- $S =$ the section modulus of the footing plan area, which is determined as

$$S = \frac{b \times d^2}{6} = \frac{(1)(2.5)^2}{6}\ 1.042\ \text{ft}^3\ [0.0295\ \text{m}^3]$$

The limiting maximum and minimum soil pressures are determined as

$$p = \frac{N}{A} \pm \frac{M}{S} = \frac{1167}{2.5} \pm \frac{5793/12}{1.042} = 467 \pm 463$$

$$= 930\ \text{psf}\ [44.5\ \text{kPa}]\ \text{maximum, and 4psf}\ [0.19\ \text{kPa}]\ \text{minimum}$$

Because the maximum stress is less than the limit of 1500 psf, vertical soil pressure is not critical for the wall.

For the horizontal force analysis, the procedure varies with different building codes. Some codes permit the addition of the two components of horizontal resistance: sliding friction on the footing bottom, and horizontal pressure on the buried structure on the low side of the wall (H in Figure 14.6). Using an average value for the friction coefficient of 0.25, the analysis is as follows:

- Total active pushing force is 454 lb [2019 N] (H_2 in Figure 14.6)
- Friction resistance = (friction factor)(load) = 0.25(1167) = 292 lb [1299 N]
- Passive resistance = 169 lb [752 N] (H in Figure 14.6)
- Total resistance = 292 + 169 = 461 lb [2051 N]

Because the total potential resistance exceeds the active force, the wall is not critical for sliding.

In most cases, designers consider the preceding analyses to be adequate for consideration of soil pressures. However, the stability of the wall is also questionable with regard to overturning. For this investigation, the loading condition is the same as that used for soil stress analysis and as shown in Figure 14.6. For overturn, the passive resistance at the low end of the wall (H in Figure 14.6) is not used. The summation of overturning and restoring moments is taken with respect to the toe of the footing (the lower-left corner in the wall section in Figure 14.6). Determination of the overturning and dead load restoring moments is shown in Table 14.2. The safety factor against overturning is determined as

$$SF = \frac{\text{restoring moment}}{\text{overturning moment}} - \frac{21,700}{9988} = 2.17$$

Overturning is usually not considered to be critical if the safety factor is at least 1.5.

A typical form for the short retaining wall is shown in Figure 14.7, in which the wall design variables are listed by letters and numbers. This figure will be used as a reference for the design of the wall and footing.

TABLE 14.2 Analysis for the Overturning Effect

	Force (lb)	Moment Arm (in.)	Moment (lb-in.)
Overturn:			
H_2	454	22	9988
Restoring Moment:			
w_1	466	18	8388
w_2	312	15	4680
w_3	311	26	8086
w_4	78	7	546
Total Restoring Moment:			21,700

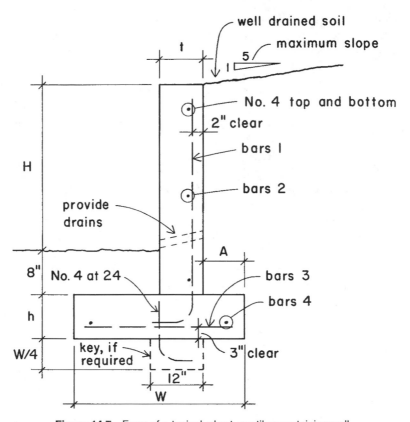

Figure 14.7 Form of a typical, short, cantilever retaining wall.

The loading condition used to analyze the stress conditions in the wall is shown in Figure 14.8. For the maximum horizontal soil pressure at the base of the wall, we determine

$$p = (30)(4.667) = 140 \text{ psf } [6.7 \text{ kPa}]$$

and the total horizontal force on the wall is, thus,

$$H_1 = \frac{(140)(4.667)}{2} = 327 \text{ lb } [1454 \text{ N}]$$

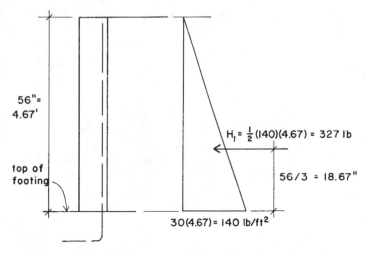

Figure 14.8 Loading for investigation of bending in the wall.

The maximum bending moment at the bottom of the wall is, thus,

$$M = (327)\left(\frac{56}{3}\right) = 6104 \text{ lb-in. [0.69 kN-m]}$$

This is adjusted for the strength method to

$$M_u = 1.6(6104) = 9766 \text{ lb-in. [1.104 kN-m]}$$

And the required resisting moment is

$$M_r = \frac{M_u}{\phi} = \frac{9766}{0.9} = 10{,}851 \text{ lb-in. [1.226 kN-m]}$$

Assuming an approximate depth for bending of 5.5 in., the tension reinforcement required for the wall (bars 1 in Figure 14.7) is

$$A_s = \frac{M_r}{f_y jd} = \frac{10{,}851}{(40{,}000)(0.9)(5.5)} = 0.055 \text{ in.}^2/\text{ft. width [116 mm}^2/\text{mm]}$$

From Table 6.6, this area may be provided by using No. 3 bars at 20-in. centers. However, the usual recommended maximum spacing is 18 in., so the choice would be for No. 3 bars at 18 in. Because the development length of the bars in the footing is quite short, they should be selected from the smaller bar sizes and should have hooked ends, as shown in Figure 14.7.

As with most wall footings, it is usually desirable to select the footing thickness to minimize the need for tension reinforcement due to bending. Thus, shear and bending in the concrete are seldom critical. For bending, the critical location is at the face of the wall, and the loading condition is as shown in Figure 14.9. The trapezoidal stress distribution produces the resultant force of 833 lb, which acts at the centroid of the trapezoidal stress block. Assuming an approximate depth of 6.5 in., the analysis is as follows:

$$M = 833(7.706) = 6419 \text{ in.-lb } [0.725 \text{ kN-m}]$$

$$M_u = 1.6(6419) = 10{,}270 \text{ in.-lb } [1.16 \text{ kN-m}]$$

$$\text{Required } M_r = \frac{10{,}270}{0.9} = 11{,}412 \text{ in.-lb } [1.29 \text{ kN-m}]$$

The required area for the bars (bars 3 in Figure 14.7) is

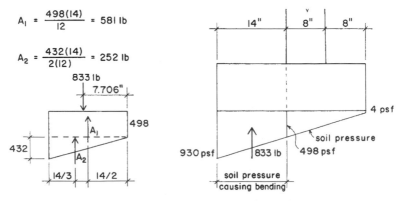

Figure 14.9 Bending investigation for the footing.

$$A_s = \frac{M_r}{f_y jd} = \frac{11,412}{(40,000)(0.9)(6.5)} = 0.049 \text{ in.}^2/\text{ft width } [103.7 \text{ mm}^2/\text{m}]$$

For ease of construction, it is desirable to have the wall and footing bars at the same spacing. The vertical bars can then be held in position for pouring by having their hooked ends wired to the footing bars. Thus, the choice for these footing bars would also be No. 3 at 18 in.

Although development is also a concern for the footing reinforcement, it is not likely to be critical so long as the bar size is small—less than No. 6 or so.

Reinforcement in the long direction of the footing (bars 4 in Figure 14.7) should be determined in the same manner as for an ordinary wall footing (see Section 13.5). A recommended minimum is for 0.15% of the footing cross section. For the 10-in.-thick and 30-in.-wide footing, this requires

$$A_s = 0.0015(300) = 0.45 \text{ in.}^2 [290 \text{ mm}^2]$$

which can be provided by three No. 4 bars with a total of 0.6 in.^{2} [387 mm^2].

Problems 14.3.A, B. Investigate a short retaining wall similar in form to that shown in Figure 14.7 for soil pressure and overturning, and select the wall reinforcement. Use = 3 ksi [20.7 MPa] and f_y = 40 ksi [276 MPa], soil weight = 100 pcf [1602 kg/m^3], and horizontal soil pressure = 30 psf/ft [4.72 kPa/m] of depth below grade. Maximum allowable vertical soil pressure is 2000 psf [95.8 kPa]. Wall dimensions in inches are as follows:

	H	W	A	h	t
A	54	36	12	12	9
B	60	48	18	14	10

14.4 ABUTMENTS

The support of some types of structures, such as arches, gables, and shells, often requires the resolution of both horizontal and vertical forces. When this resolution is accomplished entirely by the supporting foundation element, the element is described as an abutment. Figure 14.10*a* shows a simple abutment for an arch, consisting of a rectangular footing and an inclined pier. The design of such a foundation has three primary concerns as follows:

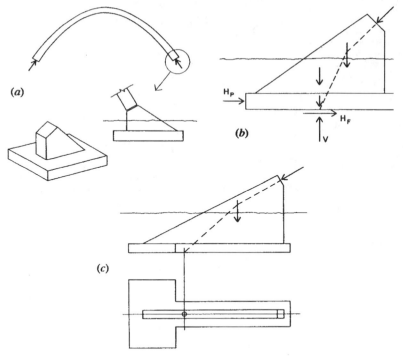

Figure 14.10   Abutments for arches.

- *Resolution of the Vertical Force.* This consists of ensuring that the vertical soil pressure does not exceed the maximum allowable value for the soil.

- *Resolution of the Horizontal Force.* If the abutment is freestanding, resolution of the horizontal force means the development of sufficient soil friction and passive horizontal pressure.

- *Resolution of the Moment Effect.* In this case, the aim is usually to keep the resultant force as close as possible to the centroid of the footing plan area. If this is truly accomplished, that is, $e = 0$, there will literally be no moment effect on the footing itself.

Figure 14.10b shows the various forces that act on an abutment, such as that shown for the arch in Figure 14.10a. The active forces consist of the load and the weights of the pier, the footing, and the soil above the footing. The reactive forces consist of the vertical soil pressure, the horizontal friction on the bottom of the footing, and the passive horizontal soil pressure against the sides of the footing and pier. The dashed line in the illustration indicates the path of the resultant of the active forces; the condition shown is the ideal one, with the path coinciding with the centroid of the footing plan area at the bottom of the footing.

If the passive horizontal pressure is ignored, the condition shown in Figure 14.10b will result in no moment effect on the bottom of the footing, and a uniform distribution of the vertical soil pressure. If the passive horizontal pressure is included in the force summation, the resultant path will move slightly to the right of the footing centroid. However, for the abutment as shown, the resultant of the passive pressure will be quite close to the bottom of the footing, so that the error is relatively small.

If the pier is tall and the load is large with respect to the pier weight, or the load is inclined at a considerable angle from the vertical, it may be necessary to locate the footing centroid at a considerable distance horizontally from the load point at the top of the pier. This could result in a footing of greatly extended length if a rectangular plan form is used. One device that is sometimes used to avoid this is a T-shape, or other form, that results in a relocation of the centroid without excessive extension of the footing. Figure 14.10c shows the use of a T-shape footing for such a condition.

When the structure being supported is symmetrical, such as an arch with its supports at the same elevation, it may be possible to resolve the horizontal force component at the support without relying on soil stresses. The basic technique for accomplishing this is to tie the two opposite supports together, as shown in Figure 14.11a, so that the horizontal force is resolved internally (within the structure) instead of externally (by the ground). If this tie is attached at the point of contact between the structure and the pier, as shown in Figure 14.11a, the net load delivered to the pier is simply a vertical force, and the pier and footing could theoretically be developed in the same manner as that for a truss or beam, without the horizontal force effect. However, because either wind or seismic loading will produce some horizontal force on the supports, the inclined pier is still the normal form for the supporting structure. The position of the footing, however, would usually be established by locating its centroid directly below the support point, as shown in the illustration.

For practical reasons, it is often necessary to locate the tie, if one is used, below the support point for the structure. If this support point is aboveground, as it usually is, the existence of the tie aboveground is quite

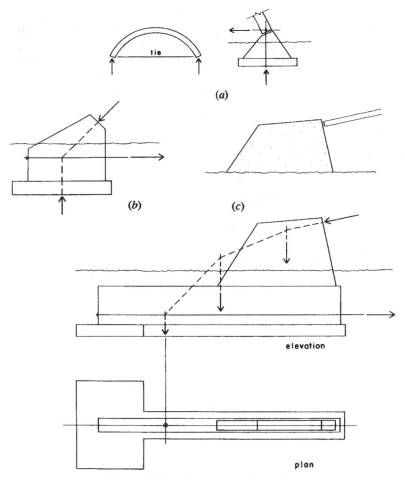

Figure 14.11   Abutments for tied arches.

likely to interfere with the use of the structure. A possible solution in this problem is to move the tie down to the pier, as shown in Figure 14.11*b*. In this case, the pier weight is added to the load to find the proper location for the footing centroid.

When the footing centroid must be moved a considerable distance from the load point, it is sometimes necessary to add another element to the abutment system. Figure 14.11*c* shows a structure in which a large grade beam has been inserted between the pier and the footing. In the example, however, it also serves to provide for the anchorage of the tie. Because of this location of the tie, the weights of both the pier and grade beam would be added to the load to find the proper location of the footing centroid. In this way, the heavy grade beam further assists the footing by helping to move the centroid closer to the pier and reducing the cantilever distance.

15

GENERAL
CONSIDERATIONS FOR
BUILDING STRUCTURES

This chapter contains discussions of some general issues relating to the design of building structures. For the most part, these concerns have not been addressed in the presentations in earlier chapters, but require consideration when dealing with whole building design situations. Application of these materials is illustrated in the design examples in Chapter 16.

15.1 CHOICE OF BUILDING CONSTRUCTION

Materials, methods, and details of building construction vary considerably on a regional basis. Many factors affect this situation, including the effects of response to climate and regional availability of construction materials. Even in a single region, differences occur between individual buildings based on styles of architectural design and techniques of builders. Nevertheless, at any given time, usually a few predominant, popular methods of construction are employed for most buildings of a given type and size. The construction methods and details shown here are reasonable, but in no way are they intended to illustrate a singular, superior style of building.

It is not possible to choose the materials and forms for a building structure without considering its integration with the general building construction. In some cases, it may also be necessary to consider the elements required for various building services, such as those for piping, electrical service, lighting, communication, roof drainage, and the heating, ventilating, and air conditioning (HVAC) systems.

For multistory buildings, it is necessary to accommodate the placement of stairs, elevators, and the vertical elements for various building services—particularly for air ducts between building levels. A major consideration for multistory buildings is the planning of the various levels so that they work when superimposed on top of each other. Bearing walls and columns must be supported from below.

Choice of both the general structural system and the various individual elements of the system is typically highly dependent on the general architectural design of the building. Ideally, the two issues—structural planning and architectural planning—are dealt with simultaneously, from preliminary design to final construction drawings.

15.2   STRUCTURAL DESIGN STANDARDS

Use of methods, procedures, and reference data for structural design is subject to the judgment of the designer. Many guides exist, but some individual selection is often required. Strong influences on choices include:

- Building code requirements, from the enforceable statutes relating to the location of the building
- Acceptable design standards as published by professional groups, such as the American Society of Civil Engineers (ASCE), referred to frequently in this book (Reference 2)
- Recommended design standards from industry organizations, such as the American Institute of Steel (AISC) and American Concrete Institute (ACI)
- The body of work from current texts and references produced by respected authors

Some reference is made to these sources in this book. However, much of the work is also simply presented in a manner familiar to the authors, based on their own experiences. If study of this subject is pursued by readers, they are sure to encounter styles and opinions that differ from those presented here. Making one's own choices in face of those conflicts is part of the progress of professional growth.

15.3 LOADS FOR STRUCTURAL DESIGN

Loads used for structural design must be derived primarily from enforceable building codes. However, the principal concern of codes is public health and safety. Performance of the structure for other concerns may not be adequately represented in the minimum requirements of the building code. Issues sometimes not included in code requirements are:

- Effects of deflection of spanning structures on nonstructural elements of the construction
- Sensations of bounciness of floors felt by building occupants
- Protection of structural elements from damage due to weather or normal usage

It is quite common for professional structural designers to experience situations in which they use their own judgment in assigning design loads. This ordinarily means using increased loads because the minimum loads required by codes must always be recognized.

Building codes currently stipulate both the load sources and the form of combinations to be used for design. The following loads are listed in the 2002 edition of the *ASCE Minimum Design Loads for Buildings and Other Structures* (Reference 2), hereinafter referred to as ASCE 2002.

- D = Dead load
- E = Earthquake-induced force
- L = Live load, except roof load
- L_r = Roof live load
- S = Snow load
- W = Load due to wind pressure

Additional special loads are listed, but these are the commonly occurring loads. The following is a description of some of these loads.

15.4 DEAD LOADS

Dead load consists of the weight of the materials of which the building is constructed, such as walls, partitions, columns, framing, floors, roofs, and ceilings. In the design of a beam or column, the dead load used must include an allowance for the weight of the structural member itself. Table 15.1,

TABLE 15.1   Weight of Building Construction

	psf^a	kPa^a
Roofs		
3-ply ready roofing (roll, composition)	1	0.05
3-ply felt and gravel	5.5	0.26
5-ply felt and gravel	6.5	0.31
Shingles: Wood	2	0.10
Asphalt	2–3	0.10–0.15
Clay tile	9–12	0.43–0.58
Concrete tile	6–10	0.29–0.48
Slate, 3 in.	10	0.48
Insulation: fiberglass batts	0.5	0.025
Foam plastic, rigid panels	1.5	0.075
Foamed concrete, mineral aggregate	2.5/in.	0.0047/mm
Wood rafters: 2 × 6 at 24 in.	1.0	0.05
2 × 8 at 24 in.	1.4	0.07
2 × 10 at 24 in.	1.7	0.08
2 × 12 at 24 in.	2.1	0.10
Steel deck, painted: 22 gage	1.6	0.08
20 gage	2.0	0.10
Skylights: Steel frame with glass	6–10	0.29–0.48
Aluminum frame with plastic	3–6	0.15–0.29
Plywood or softwood board sheathing	3.0/in.	0.0057/mm
Ceilings		
Suspended steel channels	1	0.05
Lath: Steel mesh	0.5	0.025
Gypsum board, 1/2 in.	2	0.10
Fiber tile	1	0.05
Drywall, gyspum board, 1/2 in.	2.5	0.12
Plaster: Gypsum	5	0.24
Cement	8.5	0.41
Suspended lighting and HVAC, average	3	0.15
Floors		
Hardwood, 1/2 in.	2.5	0.12
Vinyl tile	1.5	0.07
Ceramic tile: 3/4 in.	10	0.48
Thin-set	5	0.24
Fiberboard underlay, 0.625 in.	3	0.15
Carpet and pad, average	3	0.15
Timber deck	2.5/in.	0.0047/mm
Steel deck, stone concrete fill, average	35–40	1.68–1.92
Concrete slab deck, stone aggregate	12.5/in.	0.024/mm
Lightweight concrete fill	8.0/in.	0.015/mm

(continued)

TABLE 15.1 *(Continued)*

	psf[a]	kPa[a]
Floors (Continued)		
Wood joists: 2 × 8 at 16 in.	2.1	0.10
2 × 10 at 16 in.	2.6	0.13
2 × 12 at 16 in.	3.2	0.16
Walls		
2 × 4 studs at 16 in., average	2	0.10
Steel studs at 16 in., average	4	0.20
Lath. plaster—see *Ceilings*		
Drywall, gypsum board, 1/2 in.	2.5	0.10
Stucco, on paper and wire backup	10	0.48
Windows, average, frame + glazing:		
Small pane, wood or metal frame	5	0.24
Large pane, wood or metal frame	8	0.38
Increase for double glazing	2–3	0.10–0.15
Curtain wall, manufactured units	10–15	0.48–0.72
Brick veneer, 4 in., mortar joints	40	1.92
1/2 in., mastic-adhered	10	0.48
Concrete block:		
Lightweight, unreinforced, 4 in.	20	0.96
6 in.	25	1.20
8 in.	30	1.44
Heavy, reinforced, grouted, 6 in.	45	2.15
8 in.	60	2.87
12 in.	85	4.07

[a] Average weight per square foot of surface, except as noted.

Values given as /in. or /mm are to be multiplied by actual thickness of material.

which lists the weights of many construction materials, may be used in the computation of dead loads. Dead loads are due to gravity, and they result in downward vertical forces.

Dead load is generally a permanent load once the building construction is completed, unless remodeling or rearrangement of the construction occurs. Because of this permanent, longtime character, the dead load requires certain considerations in design, such as the following:

1. It is always included in design loading combinations, except for investigations of singular effects, such as deflections due to live load only.

2. Its longtime character has some special effects, creating sag and requiring reduction of design stresses in wood structures; causing long-term, continuing settlements in some soils; and producing creep effects in concrete structures.

3. It contributes some unique responses, such as the stabilizing effects that resist uplift and overturn due to wind forces.

Although weights of materials can be reasonably accurately determined, the complexity of most building construction makes the computation of dead loads possible only on an approximate basis. This adds to other factors to make design for structural behaviors a very approximate science. As in other cases, this should not be used as an excuse for sloppiness in the computational work, but it should be recognized as a fact to temper concern for high accuracy in design computations.

15.5 BUILDING CODE REQUIREMENTS FOR STRUCTURES

Structural design of buildings is most directly controlled by building codes, which are the general basis for the granting of building permits—the legal permission required for construction. Building codes (and the permit-granting process) are administered by some unit of government: city, county, or state. Most building codes, however, are based on some model code.

Model codes are more similar than different and are in turn largely derived from the same basic data and standard reference sources, including many industry standards. In the several model codes and many city, county, and state codes, however, some items reflect particular regional concerns. With respect to control of structures, all codes have materials (all essentially the same) that relate to the following issues:

1. *Minimum Required Live Loads.* All building codes have tables that provide required values to be used for live loads. Tables 15.2 and 15.3 contain some loads as specified in ASCE 2002 (Reference 2).

2. *Wind Loads.* These are highly regional in character with respect to concern for local windstorm conditions. Model codes provide data with variability on the basis of geographic zones.

3. *Seismic (Earthquake) Effects.* These are also regional, with predominant concerns in the western states. These data, including rec-

TABLE 15.2   Minimum Floor Live Loads

Building Occupancy or Use	Uniformly Distributed Load (psf)	Concentrated Load (lb)
Apartments and Hotels		
Private rooms and corridors serving them	40	
Public rooms and corridors serving them	100	
Dwellings, One-and Two-Family		
Uninhabitable attics without storage	10	
Uninhabitable attics with storage	20	
Habitable attics and sleeping rooms	30	
All other areas except stairs and balconies	40	
Office Buildings		
Offices	50	2000
Lobbies and first-floor corridors	100	2000
Corridors above first floor	80	2000
Stores		
Retail		
First floor	100	1000
Upper floors	75	1000
Wholesale, all floors	125	1000

Source: ASCE 2002 (Reference 2), used with permission of the publishers, American Society of Civil Engineers.

ommended investigations, are subject to quite frequent modification, because the area of study responds to ongoing research and experience.

4. *Load Duration.* Loads or design stresses are often modified on the basis of the time span of the load, varying from the life of the structure for dead load to a few seconds for a wind gust or a single major seismic shock. Safety factors are frequently adjusted on this basis. Some applications are illustrated in the work in the design examples in Chapter 16.

5. *Load Combinations.* These were formerly mostly left to the discretion of designers but are now quite commonly stipulated in codes, mainly because of the increasing use of ultimate strength design and the use of factored loads.

TABLE 15.3 Live Load Element Factor, K_{LL}

Element	K_{LL}
Interior columns	4
Exterior columns without cantilever slabs	4
Edge columns with cantilever slabs	3
Corner columns with cantilever slabs	2
Edge beams without cantilever slabs	2
Interior beams	2
All other members not identified above	1

Source: ASCE 2002 (Reference 2), used with permission of the publishers, American Society of Civil Engineers.

6. *Design Data for Types of Structures.* These deal with basic materials (wood, steel, concrete, masonry, etc.), specific structures (rigid frames, towers, balconies, pole structures, etc.), and special problems (foundations, retaining walls, stairs, etc.). Industry-wide standards and common practices are generally recognized, but local codes may reflect particular local experience or attitudes. Minimal structural safety is the general basis, and some specified limits may result in questionably adequate performances (bouncy floors, cracked plaster, etc.).

7. *Fire Resistance.* For the structure, there are two basic concerns, both of which produce limits for the construction. The first concern is for structural collapse or significant structural loss. The second concern is for containment of the fire to control its spread. These concerns produce limits on the choice of materials (e.g., combustible or noncombustible) and some details of the construction (cover on reinforcement in concrete, fire insulation for steel beams, etc.).

The work in the design examples in Chapter 16 is based largely on criteria from ASCE 2002 (Reference 2).

15.6 LIVE LOADS

Live loads technically include all the nonpermanent loadings that can occur, in addition to the dead loads. However, the term as commonly used usually refers only to the vertical gravity loadings on roof and floor surfaces. These loads occur in combination with the dead loads, but are generally random in character and must be dealt with as potential contributors to various loading combinations, as discussed in Section 15.8.

Roof Loads

In addition to the dead loads they support, roofs are designed for a uniformly distributed live load. The minimum specified live load accounts for general loadings that occur during construction and maintenance of the roof. For special conditions, such as heavy snowfalls, additional loadings are specified.

The minimum roof live load in psf is specified in ASCE 2002 (Reference 2) in the form of an equation, as follows:

$$L_r = 20\,R_1\,R_2 \text{ in which } 12 \le L_r \le 20$$

In the equation, R_1 is a reduction factor based on the tributary area supported by the structural member being designed (designated as A_t and quantified in ft^2), and is determined as follows:

$$R_1 = 1, \text{ for } A_t \le 200 \text{ ft}^2$$
$$= 1.2 - 0.001\,A_t, \text{ for } 200 \text{ ft}^2 < A_t < 600 \text{ ft}^2$$
$$= 0.6, \text{ for } A_t \ge 600 \text{ ft}^2$$

Reduction factor R_2 accounts for the slope of a pitched roof and is determined as follows:

$$R_2 = 1, \text{ for } F \le 4$$
$$= 1.2 - 0.05\,F, \text{ for } 4 < F < 12$$
$$= 0.6, \text{ for } F \ge 12$$

The quantity F in the equations for R_2 is the number of inches of rise per foot for a pitched roof (for example, $F = 12$ indicates a rise of 12 in 12, or an angle of 45°).

The design standard also provides data for roof surfaces that are arched or domed, and for special loadings for snow or water accumulation. Roof surfaces must also be designed for wind pressures on the roof surface, both upward and downward. A special situation that must be considered is that of a roof with a low dead load and a significant wind load that exceeds the dead load.

Although the term *flat roof* is often used, there is generally no such thing; all roofs must be designed for some water drainage. The minimum required pitch is usually $\frac{1}{4}$ in./ft, or a slope of approximately 1:50. With roof surfaces that are close to flat, a potential problem is that of *ponding*, a phenomenon in which the weight of the water on the surface causes deflection of the supporting structure, which in turn allows for more water accumulation (in a "pond"), causing more deflection, and so on, resulting in a progressive collapse condition.

Floor Live Loads

The live load on a floor represents the probable effects created by the occupancy. It includes the weights of human occupants, furniture, equipment, stored materials, and so on. All building codes provide minimum live loads to be used in the design of buildings for various occupancies. Because there is a lack of uniformity among different codes in specifying live loads, the local code should always be used. Table 15 .2 contains a sample of values for floor live loads, as given in ASCE 2002 (Reference 2) and commonly specified by building codes.

Although expressed as uniform loads, code-required values are usually established large enough to account for ordinary concentrations. For offices, parking garages, and some other occupancies, codes often require the consideration of a specified concentrated load, as well as the distributed loading. This required concentrated load is listed in Table 15.2 for the appropriate occupancies.

Where buildings are to contain heavy machinery, stored materials, or other contents of unusual weight, these must be provided for individually in the design of the structure.

When structural framing members support large areas, most codes allow some reduction in the total live load to be used for design. These reductions, in the case of roof loads, are incorporated in the formulas for roof loads given previously. The following is the method given in ASCE 2002 (Reference 2) for determining the reduction permitted for beams, trusses, or columns that support large floor areas.

The design live load on a member may be reduced in accordance with the formula

$$L = L_0 \left(0.25 + \frac{15}{\sqrt{K_{LL}A_T}} \right)$$

where:

- L = reduced design live load per square foot of area supported by the member

- L_0 = unreduced live load supported by the member
- K_{LL} = live load element factor (see Table 15.3)
- A_T = tributary area supported by the member

L shall not be less than $0.50L_0$ for members supporting one floor, and L shall not be less than $0.40L_0$ for members supporting two or more floors.

In office buildings and certain other building types, partitions may not be permanently fixed in location, but may be erected or moved from one position to another in accordance with the requirements of the occupants. In order to provide for this flexibility, it is customary to require an allowance of 15 to 20 psf, which is usually added to other dead loads.

15.7 LATERAL LOADS (WIND AND EARTHQUAKE)

As used in building design, the term *lateral load* is usually applied to the effects of wind and earthquakes because they induce horizontal forces on stationary structures. From experience and research, design criteria and methods in this area are continuously refined, with recommended practices being presented through the various model building codes.

Space limitations do not permit a complete discussion of the topic of lateral loads and design for their resistance. The following discussion summarizes some of the criteria for design in ASCE 2002 (Reference 2). Examples of application of these criteria are given in the examples of building structural design in Chapter 16. For a more extensive discussion, the reader is referred to *Simplified Building Design for Wind and Earthquake Forces* (Reference 6).

Wind

Where wind is a regional problem, local codes are often developed in response to local conditions. Complete design for wind effects on buildings includes a large number of both architectural and structural concerns. The following is a discussion of some of the requirements from ASCE 2002 (Reference 2):

Basic Wind Speed.   This is the maximum wind speed (or velocity) to be used for specific locations. It is based on recorded wind histories and adjusted for some statistical likelihood of occurrence. For the United States, recommended minimum wind speeds are taken from maps pro-

vided in the ASCE standard. As a reference point, the speeds are those recorded at the standard measuring position of 10 m (approximately 33 ft)
above the ground surface.

Wind Exposure.   This refers to the conditions of the terrain surrounding the building site. The ASCE standard uses three categories, labeled B, C, and D. Qualifications for categories are based on the form
and size of wind-shielding objects within specified distances around the
building.

Simplified Design Wind Pressure (p_s).   This is the basic reference
equivalent static pressure based on the critical wind speed, and is determined as follows:

$$p_s = \lambda\, I\, p_{S30}$$

where:

- λ = adjustment factor for building height and exposure
- I = importance factor
- p_{S30} = simplified design wind pressure for exposure B, at
  height of 30 ft, and for $I = 1.0$

The importance factor for ordinary circumstances of building occupancy is 1.0. For other buildings, factors are given for facilities that involve hazard to a large number of people, for facilities considered to be
essential during emergencies (such as windstorms), and for buildings
with hazardous contents.

The design wind pressure may be positive (inward) or negative (outward, suction) on any given surface. Both the sign and the value for the
pressure are given in the design standard. Individual building surfaces, or
parts thereof, must be designed for these pressures.

Design Methods.   Two methods are described in the Code for the application of wind pressures.

Method 1 (Simplified Procedure). This method is permitted to be used
for relatively small, low-rise buildings of simple symmetrical
shape. It is the method described here and used for the examples in
Chapter 16.

Method 2 (Analytical Procedure). This method is much more complex and is prescribed to be used for buildings that do not fit the limitations described for Method 1.

Uplift.   Uplift may occur as a general effect, involving the entire roof or even the whole building. It may also occur as a local phenomenon, such as that generated by the overturning moment on a single shear wall.

Overturning Moment.   Most codes require that the ratio of the dead load resisting moment (called the *restoring moment, stabilizing moment,* etc.) to the overturning moment be 1.5 or greater. When this is not the case, uplift effects must be resisted by anchorage capable of developing the excess overturning moment. Overturning may be a critical problem for the whole building, as in the case of relatively tall and slender tower structures. For buildings braced by individual shear walls, trussed bents, and rigid-frame bents, overturning is investigated for the individual bracing units.

Drift.   Drift refers to the horizontal deflection of the structure due to lateral loads. Code criteria for drift are usually limited to requirements for the drift of a single story (horizontal movement of one level with respect to the next above or below). As in other situations involving structural deformations, effects on the building construction must be considered; thus, the detailing of curtain walls or interior partitions may affect limits on drift.

Special Problems.   The general design criteria given in most codes are applicable to ordinary buildings. More thorough investigation is recommended (and sometimes required) for special circumstances, such as the following:

1. *Tall Buildings.* These are critical with regard to their height dimension, as well as the overall size and number of occupants inferred. Local wind speeds and unusual wind phenomena at upper elevations must be considered.
2. *Flexible Structures.* These may be affected in a variety of ways, including vibration or flutter, as well as simple magnitude of movements.
3. *Unusual Shapes.* Open structures, structures with large overhangs or other projections, and any building with a complex shape should be

carefully studied for the special wind effects that may occur. Wind-tunnel testing may be advised or even required by some codes.

Earthquakes

During an earthquake, a building is shaken up and down and back and forth. The back-and-forth (horizontal) movements are typically more violent and tend to produce major destabilizing effects on buildings; thus, structural design for earthquakes is mostly done in terms of considerations for horizontal (called *lateral*) forces. The lateral forces are actually generated by the weight of the building—or, more specifically, by the mass of the building that represents both an inertial resistance to movement and a source for kinetic energy, once the building is actually in motion. In the simplified procedures of the equivalent static force method, the building structure is considered to be loaded by a set of horizontal forces consisting of some fraction of the building weight. An analogy would be to visualize the building as being rotated vertically 90° to form a cantilever beam, with the ground as the fixed end and with a load consisting of the building weight.

In general, design for the horizontal force effects of earthquakes is quite similar to design for the horizontal force effects of wind. The same basic types of lateral bracing (shear walls, trussed bents, rigid frames, etc.) are used to resist both force effects. There are indeed some significant differences, but in the main, a system of bracing that is developed for wind bracing will most likely serve reasonably well for earthquake resistance, too.

Because of the considerably more complex criteria and procedures for the design for earthquake effects, we have chosen not to illustrate it in the examples in Chapter 16. Nevertheless, the development of elements and systems for the lateral bracing of the building in the design examples here is quite applicable in general to situations where earthquakes are a predominant concern. For structural investigation, the principal difference is in the determination of the loads and their distribution in the building. Another major difference is in the true dynamic effects, critical wind force being usually represented by a single, major, one-direction punch from a gust, while earthquakes represent rapid back-and-forth, reversing-direction actions. However, once the dynamic effects are translated into equivalent static forces, design concerns for the bracing systems are very similar, involving considerations for shear, overturning, horizontal sliding, and so on.

For a detailed explanation of earthquake effects and illustrations of the investigation by the equivalent static force method, the reader is referred to *Simplified Building Design for Wind and Earthquake Forces* (Reference 6).

15.8 LOAD COMBINATIONS

The various types of load sources, as described in the preceding section, must be individually considered for quantification. However, for design work, the possible combination of loads must also be considered. Using the appropriate combinations, the design load for individual structural elements must be determined. The first step in finding the design load is to establish the critical combinations of load for the individual element. Using ASCE 2002 (Reference 2) as a reference, the following combinations are to be considered:

- 1.4(dead load)
- 1.2(dead load) + 1.6(live load) + 0.5(roof load)
- 1.2(dead load) + 1.6(roof load) + live load or 0.8(wind load)
- 1.2(dead load) + 1.6(wind load) + (live load) + 0.5(roof load)
- 1.2(dead load) + 1.0(earthquake load) + live load + 0.2(snow load)
- 0.9(dead load) + 1.0(earthquake load) or 1.6(wind load)

15.9 DETERMINATION OF DESIGN LOADS

The following example demonstrates the process of determination of loading for individual structural elements. Additional examples are presented in the building design cases in Chapter 16.

Figure 15.1 shows the plan layout for the framed structure of a multistory building. The vertical structure consists of columns; the horizontal floor structure, of a deck-and-beam system. The repeating plan unit of 24 by 32 ft is called a *column bay*. Assuming lateral bracing of the building to be achieved by other structural elements, the columns and beams shown here will be designed for dead load and live load only.

The load to be carried by each element of the structure is defined by the unit loads for dead load and live load and the *load periphery* for the individual elements. The load periphery for an element is established by the layout and dimensions of the framing system. Referring to the labeled elements in Figure 15.1, the load peripheries are as follows:

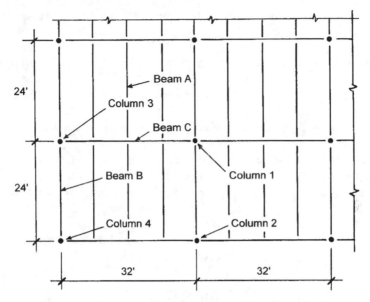

Figure 15.1 Reference for determination of distributed loads.

- Beam A: $8 \times 24 - 192$ ft^2
- Beam B: $4 \times 24 = 96$ ft^2
- Beam C: $24 \times 24 = 576$ ft^2 (Note that beam C carries only three of the four beams per bay of the system, the fourth being carried directly by the columns.)
- Column 1: $24 \times 32 = 768$ ft^2
- Column 2: $12 \times 32 = 384$ ft^2
- Column 3: $16 \times 24 = 384$ ft^2
- Column 4: $12 \times 16 = 192$ ft^2

For each of these elements, the unit dead load and unit live load from the floor are multiplied by the floor areas computed for the individual elements. Any possible live load reduction (as described in Section 15.6) is made for the individual elements based on their load periphery area.

Additional dead load for the elements consists of the dead weight of the elements themselves. For the columns and beams at the building edge, another additional dead load consists of the portion of the exterior wall construction supported by the elements. Thus, Column 2 carries an area of the exterior wall defined by the multiple of the story height times 32 ft. Column 3 carries 24 ft of wall, and Column 4 carries 28 ft of wall (12 + 16).

The column loads are determined by the indicated supported floor, to which is added the weight of the columns. For an individual story column, this would be added to loads supported above this level—from the roof and any upper levels of floor.

The loads as described are used in the defined combinations described in Section 15.8. If any of these elements are involved in the development of the lateral bracing structure, the appropriate wind or earthquake loads are also added.

Floor live loads may be reduced by the method described in Section 15.6. Reductions are based on the tributary area supported and the number of levels supported by members.

Computations of design loads using the process described here are given for the building design cases in Chapter 16.

15.10 STRUCTURAL PLANNING

Planning a structure requires the ability to perform two major tasks. The first is the logical arranging of the structure itself, regarding its geometric form, its actual dimensions and proportions, and the ordering of the elements for basic stability and reasonable interaction. All of these issues must be faced, whether the building is simple or complex, small or large, of ordinary construction or totally unique. Spanning beams must be supported and have depths adequate for the spans, horizontal thrusts of arches must be resolved, columns above should be centered over columns below, and so on.

The second major task in structural planning is the development of the relationships between the structure and the building in general. The building plan must be "seen" as a structural plan. The two may not be quite the same, but they must fit together. "Seeing" the structural plan (or possibly alternative plans) inherent in a particular architectural plan is a major task for designers of building structures.

Hopefully, architectural planning and structural planning are done interactively, not one after the other. The more the architect knows about the structural problems and the structural designer (if another person) knows about architectural problems, the more likely it is that an interactive design development may occur.

Although each individual building offers a unique situation if all of the variables are considered, the majority of building design problems are highly repetitious. The problems usually have many alternative solutions, each with its own set of pluses and minuses in terms of various points of comparison. Choice of the final design involves the comparative evaluation of known alternatives and the eventual selection of one.

The word *selection* may seem to imply that all the possible solutions are known in advance, not allowing for the possibility of a new solution. The more common the problem, the more this may be virtually true. However, the continual advance of science and technology and the fertile imagination of designers make new solutions an ever-present possibility, even for the most common problems. When the problem is truly a new one, in terms of a new building use, a jump in scale, or a new performance situation, there is a real need for innovation. Usually, however, when new solutions to old problems are presented, their merits must be compared to established previous solutions in order to justify them. In its broadest context, the selection process includes the consideration of all possible alternatives: those well known, those new and unproven, and those only imagined.

15.11 BUILDING SYSTEMS INTEGRATION

Good structural design requires integration of the structure into the whole physical system of the building. It is necessary to realize the potential influences of structural design decisions on the general architectural design and on the development of the systems for power, lighting, thermal control, ventilation, water supply, waste handling, vertical transportation, firefighting, and so on. The most popular structural systems have become so, in many cases, largely because of their ability to accommodate the other subsystems of the building and to facilitate popular architectural forms and details.

15.12 ECONOMICS

Dealing with dollar cost is a very difficult, but necessary, part of structural design. For the structure itself, the bottom-line cost is the delivered cost of the finished structure, usually measured in units of dollars per square foot of the building. For individual components, such as a single wall, units may be used in other forms. The individual cost factors or components, such as cost of materials, labor, transportation, installation, testing, and inspection, must be aggregated to produce a single unit cost for the entire structure.

Designing for control of the cost of the structure is only one aspect of the design problem, however. The more meaningful cost is that for the entire building construction. It is possible that certain cost-saving efforts applied to the structure may result in increases of cost of other parts of the construction. A common example is that of the floor structure for multistory buildings. Efficiency of floor beams occurs with the generous provision of beam depth in proportion to the span. However, adding inches to beam depths, with the unchanging need for dimensions required for floor and ceiling construction and installation of ducts and lighting elements, means increasing the floor-to-floor distance and the overall height of the building. The resulting increases in cost for the added building skin, interior walls, elevators, piping, ducts, stairs, and so on, may well offset the small savings in cost of the beams. The really effective cost-reducing structure is often one that produces major savings of nonstructural costs, in some cases at the expense of less structural efficiency.

Real costs can be determined only by those who deliver the completed construction. Estimates of cost are most reliable in the form of actual offers or bids for the construction work. The further the cost estimator is from the actual requirement to deliver the goods, the more speculative the estimate. Designers, unless they are in the actual employ of the builder, must base any cost estimates on educated guesswork deriving from some comparison with similar work recently done in the same region. This kind of guessing must be adjusted for the most recent developments in terms of the local markets, the competitiveness of builders and suppliers, and the general state of the economy. Then the four best guesses are placed in a hat and one is drawn out.

Serious cost estimating requires a lot of training and experience and an ongoing source of reliable, timely information. For major projects, various sources are available, in the form of publications or computer databases.

The following are some general rules for efforts that can be made in the structural design work in order to have an overall cost-saving attitude:

1. Reduction of material volume is usually a means of reducing cost. However, unit prices for different grades must be noted. Higher grades of steel or wood may be proportionally more expensive than the higher stress values they represent; more volume of cheaper material may be less expensive.
2. Use of standard, commonly stocked products is usually a cost savings because special sizes or shapes may be premium-priced.

Wood two-by-three studs may be higher in price than two-by-four studs because the two-by-four is so widely used and bought in large quantities.

3. Reduction in the complexity of systems is usually a cost savings. Simplicity in purchasing, handling, managing of inventory, and so on will be reflected in lower bids as builders anticipate simpler tasks. Use of the fewest number of different grades of materials, sizes of fasteners, and other such variables is as important as the fewest number of different parts. This is especially true for any assemblage done on the building site; large inventories may not be a problem in a factory, but usually are on a restricted site.

4. Cost reduction is usually achieved when materials, products, and construction methods are highly familiar to local builders and construction workers. If real alternatives exist, choice of the "usual" one is the best course.

5. Do not guess at cost factors; use real experience—yours or others. Costs vary locally, by job size and over time. Keep up to date with cost information.

6. In general, labor cost is greater than material cost. Labor for building forms, installing reinforcement, pouring, and finishing concrete surfaces is *the* major cost factor for site-poured concrete. Savings in these areas are much more significant than saving of material volume.

7. For buildings of an investment nature, time is money. Speed of construction may be a major advantage. However, getting the structure up fast is not a true advantage unless the other aspects of the construction can take advantage of the time gained.

16

BUILDING STRUCTURES: DESIGN EXAMPLES

This chapter presents examples of the design of structural systems for buildings. The building used for the examples have been chosen to create a range of situations in order to demonstrate the use of various structural components. Design of individual elements of the structural systems is largely based on materials presented in the earlier chapters. To conserve space, reference is sometimes made to computations demonstrated in the earlier chapters. Computations shown here are presented in a condensed form, without extensive explanations—the form commonly used by design engineers for a record of their design development.

Of primary concern in this chapter are the presentation of the design process for whole systems and consideration of the many factors that influence design decisions. Many of these factors do not relate to structural behaviors, but, nevertheless, they have significant influence on the final form and details of the structure.

The work here is profusely illustrated, with many cut section details to help the reader to understand the construction. While these details are mostly executed in a form used for contract documents (working drawings), their purpose here is primarily pictorial; thus, they are sometimes not as complete as required for contract work. The works shown here should not be considered as models for contract drawings.

16.1 BUILDING ONE: GENERAL CONSIDERATIONS

Building One is a split-level, single-family house, a common sight in small-town and suburban America (see Figure 16.1). The house itself is primarily executed with light wood frame construction, the most popular form for this type of building. As certain as the use of wood may be for the aboveground construction, however, it is even more certain that the ground-contacting portions of the construction will be executed with concrete or masonry construction. While some details of the supported wood structure are shown here, the principal attention is given to the supporting construction for the building and some of the site construction, shown here as using sitecast concrete.

Most houses like this are not built as single contracted buildings, but rather are part of a development project with many houses built from the same basic plans. Specific details of the construction relate to the local conditions and to the practical concerns of builders to achieve economical production. In most cases, however, construction methods are held to the simplest forms, permitting the least use of highly skilled labor or complex equipment at the building site. A high priority for the development of construction details is a desire for the easiest, fastest, lowest-cost, and least craft-dependent forms of construction, at least for anything that must be accomplished at the building site.

16.2 BUILDING ONE: SUPPORT AND SITE STRUCTURES

Figure 16.2 shows some possible forms for common elements of the supporting structure. The letters on the details refer to locations indicated on the building section in Figure 16.1.

Detail C in Figure 16.2 shows the support of a wood stud wall on top of a concrete wall. A common connecting device for this situation is the steel anchor bolt, which is cast into the top of the concrete wall. This must be located with some precision in order to bolt down the sill for the stud wall. The anchor bolts serve to hold the wood frame in position during construction. However, they may also be required to transfer lateral and uplift loads to the foundations due to wind or earthquake forces.

BUILDING 1
1 – Living Room
2 – Dining Room
3 – Kitchen
4 – Bedroom
5 – Family Room
6 – Garage

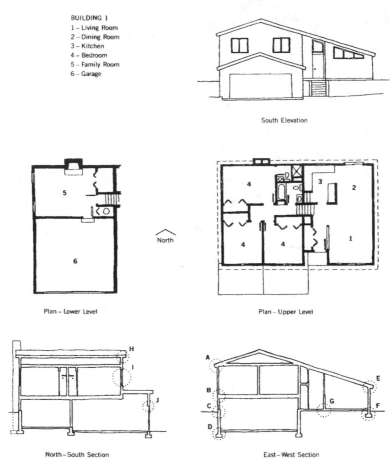

South Elevation

North

Plan – Lower Level   Plan – Upper Level

North–South Section   East–West Section

Figure 16.1 Building One: general form.

The concrete wall is shown supported on a simple strip footing in detail D of Figure 16.2. For the loads from this small, light building, this footing is likely to be only a small amount wider than the wall and likely to be executed with no reinforcement. However, there is a special problem for this particular wall that may require a different form of footing; this is discussed later.

Detail D also shows the edge of the concrete floor slab for the garage. This is most likely not actually connected to the wall, but simply floating on a prepared base on top of the ground. For passenger-car wheel loads,

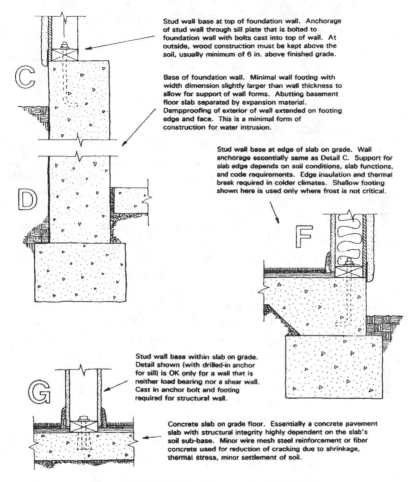

Stud wall base at top of foundation wall. Anchorage of stud wall through sill plate that is bolted to foundation wall with bolts cast into top of wall. At outside, wood construction must be kept above the soil, usually minimum of 6 in. above finished grade.

Base of foundation wall. Minimal wall footing with width dimension slightly larger than wall thickness to allow for support of wall forms. Abutting basement floor slab separated by expansion material. Dampproofing of exterior of wall extended on footing edge and face. This is a minimal form of construction for water intrusion.

Stud wall base at edge of slab on grade. Wall anchorage essentially same as Detail C. Support for slab edge depends on soil conditions, slab functions, and code requirements. Edge insulation and thermal break required in colder climates. Shallow footing shown here is used only where frost is not critical.

Stud wall base within slab on grade. Detail shown (with drilled-in anchor for sill) is OK only for a wall that is neither load bearing nor a shear wall. Cast in anchor bolt and footing required for structural wall.

Concrete slab on grade floor. Essentially a concrete pavement slab with structural integrity highly dependent on the slab's soil sub-base. Minor wire mesh steel reinforcement or fiber concrete used for reduction of cracking due to shrinkage, thermal stress, minor settlement of soil.

Figure 16.2 Building One: construction details.

the minimum 3.5-in.-thick slab is most likely adequate here. With the size of this garage, it is probably possible to pour this slab in a single unit, with separating joints occurring only at the surrounding walls.

Details F and G show some conditions for the concrete slab floor at the living room level. Detail F shows the use of a shallow foundation at the building edge, a common form used in mild climates where protection from frost action is not a concern. In cold climates, the footing must be placed considerably below the exterior surface of the ground, which is typically achieved by introducing a wall between the slab edge above and the footing below (similar to detail 8 in Figure 16.5).

Detail G in Figure 16.2 shows an interior partition wall supported directly on the floor slab. In this case, the wall sill is anchored by devices installed after the slab is cast. This is a practical solution and acceptable in most cases for interior walls that are not required to function as shear walls. Locating required anchor bolts precisely and somehow holding them in midair while concrete is poured around them is not easy, so this detail is truly a simplification of the building process.

Figure 16.3 shows a site plan that indicates a sloping lot. While changes of the ground profile can be made within the site, a common situation is the need to meet existing ground contours at the lot edge. If lowering, raising, or simply leveling out the site is required, some special construction may be necessary at the lot edges to protect the adjoining properties.

Another concern for the site is the relation between the original ground surface and the constructed surfaces. The term *original surface,* as used here, refers to an undisturbed condition before any constructed filling to raise the surface level, whether the fill is part of the current work or from some previous activity. Cutting down of the site for construction may expose materials that need consideration. On the other hand, raising finished surfaces above the original grade may require special consideration for the support of construction placed on top of the fill.

Figure 16.4 shows structural plans for the two levels of the building that sit directly on the ground. Various details of this construction and some of the site construction elements are shown in Figures 16.4 and 16.5. The following discussion relates to items shown in Figures 16.4 and 16.5:

Lower-Level Floor

The building section in Figure 16.1 and the plan of the lower level in Figure 16.4 show that the floor changes level between the garage and the family room (detail 6). The easiest way to achieve this is simply to cast a strip grade beam/footing and to cast the two floor slabs separately. This footing/wall also supports the dividing wall in this case.

The garage floor continues out to become the driveway at the entrance. The driveway slab and the garage floor slab are essentially similar, but there should be a separation joint here and some support for the slab edges (detail 5). The grade beam shown here is merely an extension of the wall-supporting construction at the edges of the door opening.

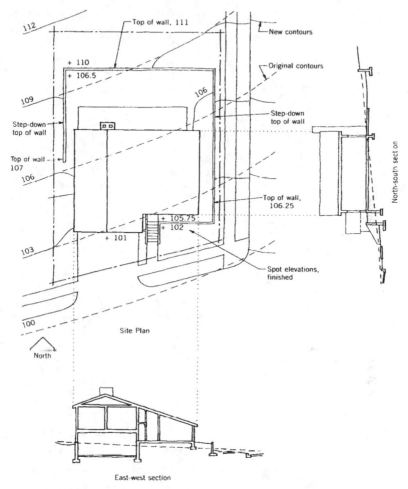

Figure 16.3   Building One: site plan and sections.

Laterally Unsupported Wall

The site plan shows that the finish ground surface slopes along the west side of the building. At the rear of the house, the level approaches that of the middle (living room) level, while at the front, it approaches the level of the garage floor. Although this foundation wall serves to form a basement space (or half basement), it does not have the usual lateral support

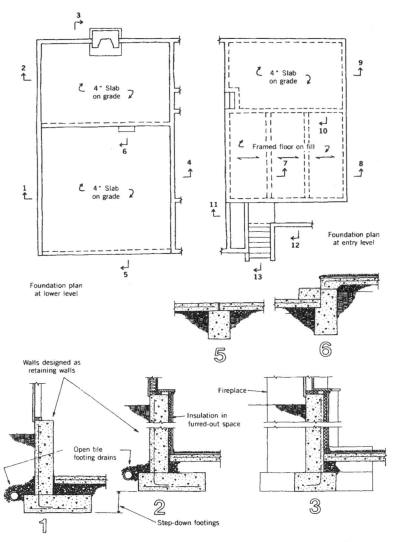

Figure 16.4 Building One: foundation and site construction details.

of a full basement wall. The bottom of the wall is adequately braced by the edge of the concrete floor slab, but its top merely supports the wood stud wall. If balloon framing (with continuous studs) is used, it may be possible to use the wall studs for lateral support. However, it is probably

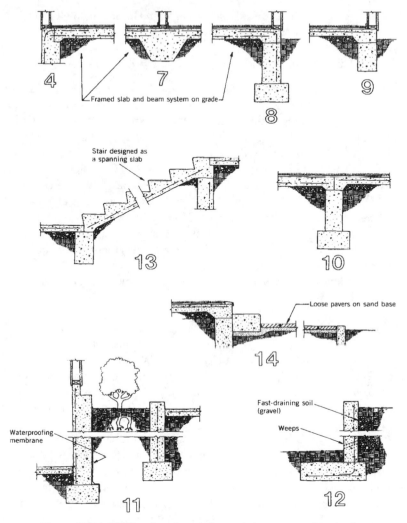

Figure 16.5   Building One: foundation and site construction details.

better to build this wall as a cantilever retaining wall, with the footing as shown in details 1 and 2 in Figure 16.4.

At the rear of the house, however, this wall can span the short horizontal distance from the side walls to the fireplace, so a simple wall footing is adequate. The wall between the two levels at the center of the house also has a difference of level on two sides, but should be adequately braced by other construction (detail 4 in Figure 16.5).

Framed Floor on Grade

The floor plan for the middle level in Figure 16.4 and details 4, 7, and 8 in Figure 16.5 show a special type of sitecast concrete construction. This consists of a spanning structure cast directly on soil materials, commonly called a *framed floor on grade*. The usual reason for using such a structure is lack of confidence in the supporting material, regarding future settlement. This is typically due to the condition that exists here, in which the finished level is some distance above the original undisturbed ground surface. Regardless of the quality of work for compaction of this fill, some settlement must be anticipated. This slight movement must be accepted for exterior construction (drives, walks, patios, etc.), but it is a different problem for building floors—especially ones that support walls or other interior construction.

Depending on circumstances, this structure may be executed in a number of ways, as follows:

One-Way Slab.   If supporting walls or grade beams are relatively close together, a simple one-way slab may be used. This consists of simply adding some reinforcing bars to the usual slab on grade construction. However, if the span is more than 8 ft or so, the slab must be thickened and may become unfeasible. In this case, the shortest span is about 18 ft, which would require a very thick slab even though the live load is low.

Two-Way Slab.   If supporting edge structures define rectangular areas of approximately square shape, it may be possible to use a two-way spanning slab. This would allow the use of a slightly thinner slab, in comparison to the one-way spanning slab. This is an option for our case because the bay size is close to square (18 ft by 22 ft, approximately), but something thicker than the 3.5-in.-thick slab would be required.

Precast Spanning Units.   If a considerable area of floor is involved, a simple solution may be to use precast floor units. For this small project, however, it is not necessary and probably not practical.

Post-Tensioned, Prestressed Slab.   This is sometimes a solution for larger projects but is much more complicated and expensive than other alternatives here.

One-Way Spanning, Slab and Beam System. This amounts basically to simply casting the usual slab on grade, with the addition of some extra reinforcement and the creation of beams by trenching slightly below the general surface developed for the bottom of the slab. This trenching can also be used (as it is more frequently) to create a strip footing for a wall, as shown in detail 7 in Figure 16.5.

The slab and beam system on grade may be designed in the same manner as the usual spanning system for floors or roofs. Some details for the system are shown in Figure 16.6.

The three-span slab is reinforced with three sets of bars, designated *A, B,* and *C* in the figure. These are placed and used as follows:

1. Set *A* is the lowest bars, which are supported to keep them above the subgrade material. These provide for positive moment bending (tension in the bottom at midspan) and in this case would be provided as continuous bars, approximately 21 ft long. If this is an impractical length, they could be spliced at one of the beams.

2. Set *B* is placed on top of the lower set, at right angles, and provides shrinkage reinforcement. Actually, set *A,* as continuous bars, also provides this function.

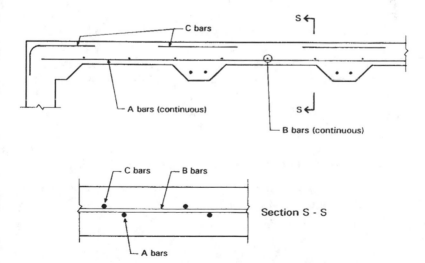

Figure 16.6 Building One: details of the framed construction on grade.

3. Set C is the top layer, providing for negative moment bending (tension in the top over the interior supports). They are shown here as cut off in the center of the spans, but it would actually probably be easier for installation, and cost-effective, if they were also made continuous.

As mentioned previously, most of the wall construction shown in the sections could be achieved with masonry (most likely concrete masonry units [CMUs]) and frequently is. This applies to construction for both the building and site structures.

16.3 BUILDING TWO: GENERAL CONSIDERATIONS

Building Two consists of a one-story, box-shaped building that is intended for commercial occupancy. For maximum flexibility in terms of interior rearrangements, it is desired to have a clear-span roof structure. Figure 16.7 shows a scheme for the building that uses CMU walls and steel, open web joists for the roof spanning structure. The following data are used for design:

- Roof live load = 20 psf [0.96 kPa] (reducible)
- Design wind pressure, ASCE Method 1
- CMUs are medium-weight units

The general profile of the building is shown in Figure 16.7c, which indicates use of a low slope roof, a flat ceiling, and a short parapet at the exterior walls. The general form of the construction is shown in Figure 16.7d. Lateral bracing is achieved with perimeter shear walls, which take the form of individual, vertically cantilevered piers. Design for lateral loads is discussed in Section 16.5. The following section presents a discussion of the design for gravity loads.

16.4 BUILDING TWO: DESIGN FOR GRAVITY LOADS

The exterior masonry walls serve as bearing walls to support the roof construction. For the dead weight of the roof construction, the following is determined (see Table 15.1):

- Three-ply felt and gravel roofing   = 5.5 psf
- Foamed concrete insulation fill, 4 in. average = 10.0

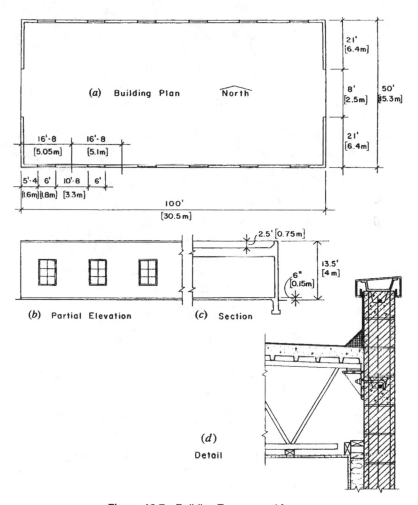

Figure 16.7   Building Two: general form.

- Formed sheet steel deck, 20 gage   = 2.0
- Steel joists, from supplier's catalog   = 12.0
- Ceiling: wood nailers and blocking   = 1.0
- Gypsum drywall   = 2.5
- Lighting, HVAC, etc.   = 3.0
- Total roof dead load   = 36.0 psf [1.72 kPa]

For the walls on the long sides of the building, which support the spanning joists, the uniformly distributed bearing load on the top of the wall is determined as follows:

- Roof dead load = (50 ft span/2)(36 psf) = 900 lb/ft
- Roof live load = (25 ft)(20 psf) = 500 lb/ft (not reduced)

Estimating the average weight of the CMU wall at 60 psf of wall surface, the weight of the wall at its bottom is:

- Wall load = (13.5 ft height)(60 psf) = 810 lb/ft
- Wall live load = 500 lb/ft
- Total dead load = 810 + 900 = 1710 lb/ft
- Factored W_u = 1.2(1710) + 1.6(500) = 2852 lb/ft [41.6 kN/m]

Assuming a nominal 8-in.-thick CMU wall, the average stress at the bottom of the wall is

$$f_b = \frac{N}{A} = \frac{2852}{7.5 \times 12} = 32 \text{ psi}$$

We will not present the design of the masonry wall, but this is a low stress, even for an *unreinforced masonry wall*.

Figure 16.8 shows the use of a short foundation wall and a footing for support of the building wall. This wall will ordinarily be formed as shown, with all voids filled with concrete. Additionally, some horizontal reinforcement should be placed at the top and bottom of the wall, permitting the wall to act as a grade beam to distribute the spaced pier loads to the footing, as shown in Figure 16.9. With the service load as computed, the footing may be quite minimal, depending on the allowable soil pressure. An unreinforced footing may quite likely be adequate.

The short foundation wall could also be made with sitecast concrete, as shown in Figure 16.10*b*. Where frost is not a problem, it is common practice to use the single grade beam foundation without a separate footing, as shown in Figure 16.10*a*.

A consideration for the building wall design is the manner in which the roof joists are supported. If the parapet wall is used and the joists are supported as shown in Figure 16.7, the roof load will develop considerable bending in the top of the wall due to the eccentricity of the load. A

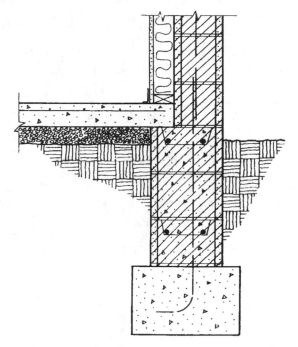

Figure 16.8   Building Two: masonry foundation wall.

modification of the construction that may serve to eliminate this bending in the wall is shown in Figure 16.11. In this case, the joists bear on top of the wall in simple direct bearing.

The masonry walls also serve as shear walls for lateral loads and must be designed for the critical load combinations described in Section 15.8. General considerations for the lateral load design are discussed in the next section.

16.5   BUILDING TWO: DESIGN FOR LATERAL LOADS

Design for the lateral effects of wind or earthquakes begins with decisions about the general form of the lateral bracing system. For this building, which consists only of a perimeter wall structure supporting a near-flat roof, the simplest system is one using perimeter shear walls and a horizontal diaphragm roof.

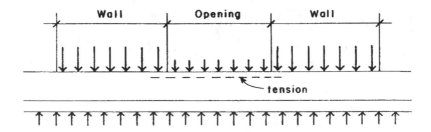

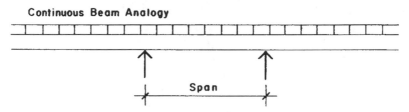

Figure 16.9   Building Two: spanning action of the foundations.

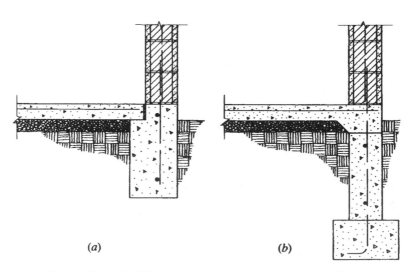

Figure 16.10   Building Two: concrete options for the foundations.

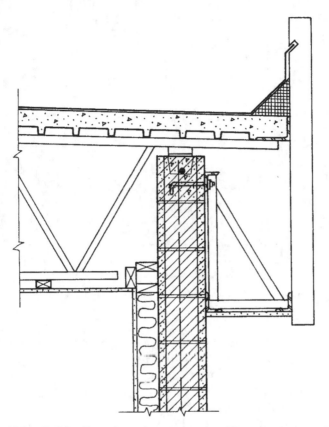

Figure 16.11 Building Two: alternative construction. For comparison see Figure 16.7.

Depending on the details of the wall construction, the walls may serve as lateral bracing for forces in their own planes by one of two means as shown in Figure 16.12. In Figure 16.12*a,* the walls are shown as consisting of individual piers acting as cantilevers with fixed bases. This action is developed by having the intervening construction between piers constituted with a break in the continuity of the wall. This could be achieved with control joints in the masonry or by a change to another form of construction between the piers.

Another possible form of action for the walls is shown in Figure 16.12*b.* In this case, the wall is considered to act as a continuous, rigid frame bent. The deep, stiff, continuous strips above and below the open-

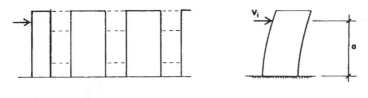

(*a*) **Wall as linked, isolated piers**

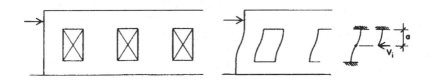

(*b*) **Continuous, pierced wall**

Figure 16.12   Building Two: functioning of the shear walls.

ings are considered to be nonflexing elements, and the wall piers between the windows, thus, act as columns with fixed tops and bottoms. If the masonry is built with reinforced construction (see Figure 10.17) and the portions above and below the windows are indeed quite deep, this is a reasonable assumption for the wall action.

For the wall action shown in Figure 16.12*a,* the individual wall piers may be designed as independent, freestanding shear walls. This design involves the following considerations:

Horizontal Shear in the Wall. A factored load is determined for the shear force and is compared to the factored resistance of the wall. The magnitude of the shear force may determine the materials, form, thickness, and reinforcement details for the wall.

Cantilever Bending of the Wall. Depending on the wall form and construction, bending may be resisted by the entire wall cross section or may be developed essentially by the two opposed wall ends acting like the flanges of an I-beam. With reinforced CMU construction, the latter is usually assumed, with the end "columns" acting in opposed tension and compression.

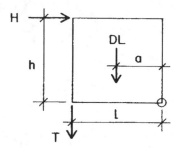

Figure 16.13 Building Two: Determination of stability and tiedown requirements for a shear wall.

Anchorage for Overturn. The basis for this is shown in Figure 16.13.

Transfer of Forces Between Elements. Forces must be transferred from the roof to the walls, and from the walls to the foundations. Details of the construction must be developed to achieve these force transfers. The wall base may need to develop a tiedown anchorage if the dead weight of the wall and the supported construction is not sufficient to resist overturn. Ordinary doweling of the vertical wall reinforcement frequently achieves this anchorage.

For any form of construction, some minimum requirements usually establish a base level of structural capability, once that form of construction is chosen. For structural masonry, code requirements for the units, mortar, and some details of the construction will define this minimum construction. In many applications, for buildings of modest size, the structural capacity of this minimum construction will be adequate. Such is essentially the case for this building.

Wind Forces on the Bracing System

The horizontal wind force on the north and south walls of the building is shown in Figure 16.14. This force is generated by a combination of positive pressure (inward, direct push) on the windward side and negative pressure (outward, suction) on the lee side of the building. The pressures shown in Case 1 in Figure 16.14 are obtained from data in the AISC 2002 (Reference 2) chapter on wind loads (see discussion in Section 15.7). These criteria provide for two zones of pressure: a general one and a small special increased pressure area at one end. The values of pressure

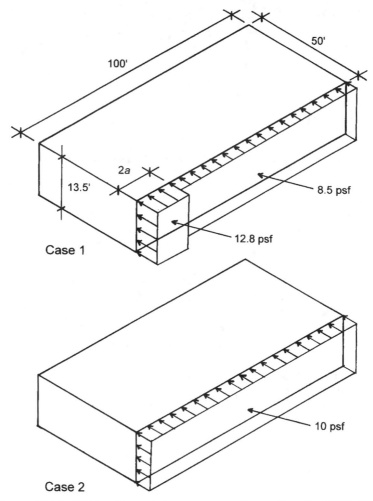

Figure 16.14 Building Two: wind pressure on the south wall, ASCE 2002 (Reference 2).

shown in Figure 16.14 are derived by considering a critical wind veloc-ity of 90 MPH and an exposure condition B, as described in the standard.

The range for the increased pressure in Case 1 is defined by the di-mension a and the height of the windward wall. The value of a is estab-lished as 10% of the least plan dimension of the building or 40% of the wall height, whichever is smaller, but not less than 3 ft. For this example, *a* is determined as 10% of 50, or 5 ft. The distance for the pressure of 12.8 psf in Case 1 is, thus, $2(a) = 10$ ft.

The design standard also requires that the bracing system be designed for a minimum pressure of 10 psf on the entire wall. This sets up two cases (Case 1 and Case 2 in Figure 16.14) that must be considered. Because the concern for the design is the generation of the maximum effect on the roof diaphragm and the end shear walls, the critical conditions may be determined by considering the development of the end reaction forces and maximum shear for an analogous beam subjected to the two loadings. This analysis is shown in Figure 16.15, from which it is apparent that the critical load for the end shear wall and the maximum effect in the roof diaphragm is derived from Case 2 in Figure 16.14.

Figure 16.16 shows the basis for determination of the total force effects on the roof diaphragm. Uplift on the roof must also be considered, but the effects shown in Figure 16.16 are the usual basis for the design of

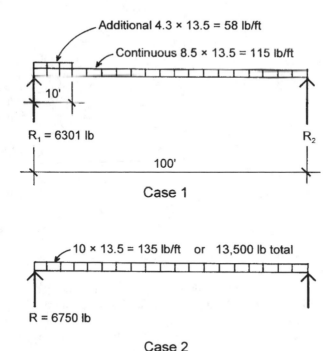

Figure 16.15   Building Two: resultant wind forces on the end shear walls.

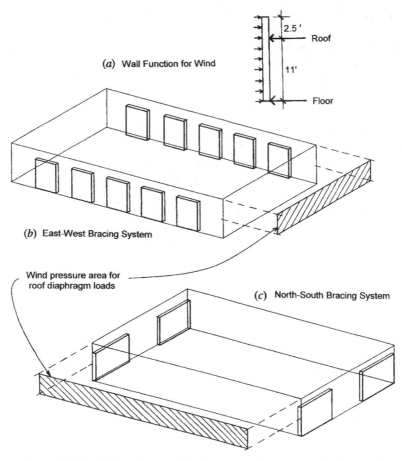

(a) Wall Function for Wind

2.5'
Roof
11'
Floor

(b) East-West Bracing System

Wind pressure area for
roof diaphragm loads

(c) North-South Bracing System

Figure 16.16 Building Two: wall functions and wind pressure development.

the roof deck for diaphragm action and for the determination of the lateral loads for the perimeter bracing.

The walls must also act as vertically spanning elements in resisting the direct wind pressure on their surfaces. As shown in Figure 16.16a, the walls span from floor to roof, acting as a beam with a cantilevered end. Some of the wind load on the wall goes directly into the edge of the floor slab and is not delivered to the edge of the roof diaphragm. Thus, the wind load to be used for the design of the roof diaphragm and its bracing walls is that shown in Figure 16.16*b* (for east or west direction wind), or that shown in Figure 16.16*c* (for north or south direction wind).

The spanning action of the walls results in bending in the walls, which must be combined with effects of gravity loads. For very tall walls or for high wind pressures, this combined action may be a critical concern. Additional bending may develop if the construction is as shown in Figure 16.7.

Considering the wind on the north and south walls and assuming a wall action as shown in Figure 16.6*a*, the north-south wind force delivered to the roof edge is determined as

$$\text{Total } W = (10 \text{ psf})(100 \times 13.5) = 13,500 \text{ lb } [60\text{kN}]$$

$$\text{Roof edge } W = 13,500 \times \frac{6.75}{11} = 8284 \text{ lb } [37.2 \text{ kN}]$$

In resisting this load, the roof functions as a spanning member supported by the shear walls at the east and west ends of the building. The investigation of the diaphragm as a 100-ft simple span beam with a uniformly distributed loading is shown in Figure 16.17.

The end reaction, which is also the maximum shear in the diaphragm and the load to the end shear walls, is found as

$$R = V = \frac{8284}{2} = 4142 \text{ lb } [18.6 \text{ kN}]$$

These forces are distributed to the individual wall piers in proportion to their number and relative stiffness. Determined in this manner, the loadings for the individual piers on all sides of the building are as shown in Figure 16.18.

The overturning moments on the wall piers also create effects on the wall foundations. Thus, the net effect of the lateral overturning moment minus the restoring moment due to dead load must be resisted by both the wall and its foundation. If a net overturn exists, some value must be determined for the anchorage force T in Figure 16.18, and the details of the construction must be developed for this effect.

A net overturning moment will develop the effect shown in Figure 16.19 on the continuous foundation wall or grade beam. Investigation for this effect will determine the need for top and bottom reinforcement in the wall/grade beam, as shown in Figures 16.8 and 16.10.

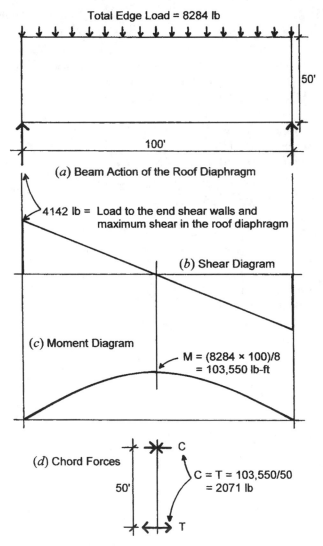

Total Edge Load = 8284 lb

50'

100'

(*a*) Beam Action of the Roof Diaphragm

4142 lb = Load to the end shear walls and
maximum shear in the roof diaphragm

(*b*) Shear Diagram

(*c*) Moment Diagram

M = (8284 × 100)/8
= 103,550 lb-ft

(*d*) Chord Forces

C

C = T = 103,550/50
= 2071 lb

50'

T

Figure 16.17   Building Two: spanning functions of the roof diaphragm.

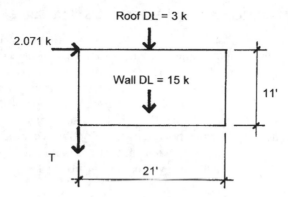

Figure 16.18 Building Two: functions of the end shear wall.

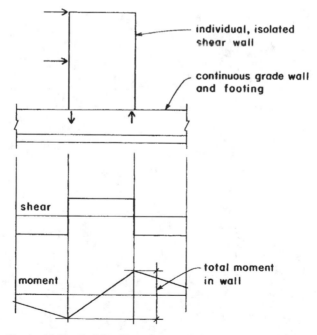

Figure 16.19 Building Two: shear wall on a continuous footing.

16.6   BUILDING TWO: ALTERNATIVE STRUCTURE

Building Two could be constructed with a great variety of materials and
systems, without major alteration of the basic plan or the general exterior
appearance of the building. A common alternative would be one using a
frame structure of wood or steel, possibly with a masonry veneer wall.
Figure 16.20 shows a possibility for a system employing walls of precast
concrete. For the plan as shown in Figure 16.7, the north and south walls
would be developed with individual 16-ft-by-8-in.-wide panels, each
with a window opening, as shown in the partial elevation in Figure 16.20.

This wall system could be developed to function in a number of ways,
such as:

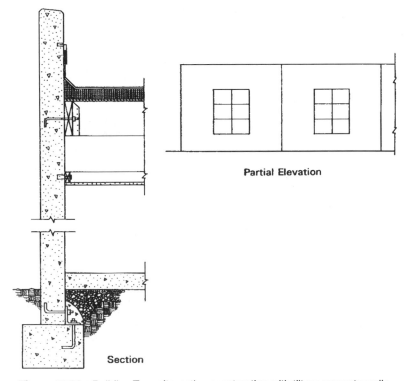

Partial Elevation

Section

Figure 16.20   Building Two: alternative construction with tilt-up concrete walls.

1. Fully structural units, providing resistance to gravity and lateral loadings

2. Shear walls only, with gravity loads supported by a frame structure to which the wall panels are attached

3. Architectural (nonstructural) elements serving no building structural purpose other than spanning vertically for wind pressures and supporting themselves for gravity loading—in other words, a curtain wall system

The system shown in Figure 16.20 could be achieved by the tilt-up method discussed in Sections 1.9 and 12.5. This is only one means for precasting, and the wall units may also be cast in a factory and transported to the site.

Another option for Building Two is the use of sitecast concrete walls. This is quite expensive, so most likely would not be chosen, unless it is a major factor in the architectural design. This is the case for the building in the next section.

16.7 BUILDING THREE: ALTERNATIVE STRUCTURE ONE

Figure 16.21 shows the general form and some details for a small, one-story building for a library.

A major feature of the design is the use of exposed elements of the concrete structure. Solid portions of the exterior walls consist of sitecast concrete, with the exterior surfaces exposed to view. The clear-span roof structure uses large, precast concrete units, with the underside exposed to view. The T-shaped roof elements extend over the concrete bearing walls to form an overhanging roof edge on the long sides of the building. The placement of the T units at the building ends also results in an overhanging roof edge.

Although both the roof and the walls are concrete, the processes for their production are quite different, so a matching of the concrete materials and finishes is not possible. It would probably be best to develop the finishes in deliberately contrasting ways in order to further differentiate the two materials. The underside of the T units may be simply cleaned and painted, while the exposed surfaces of the walls might receive a sandblasting or other treatment to produce a roughened surface, exposing the coarse aggregate materials.

In a warm climate, the interior surfaces of the exterior walls may also

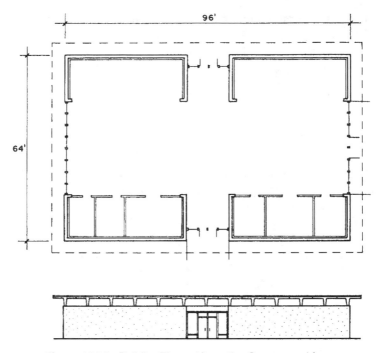

Figure 16.21   Building Three, Alternative One: general form.

be exposed. However, in cold climates it is advisable to add some insulation because the concrete itself is not very insulative and would produce a clammy interior condition in cold seasons. One possible solution
is to use a furred-out space on the interior with a gypsum drywall surface,
creating a void space for insulation.

The T-shaped units are commercial products and would be supplied
by a manufacturer who may also perform the site installation. Structural
design of the units would be done by the manufacturer's engineers or
consultants. Some dimensions and special details for the units would be
adjusted to the specific usage conditions, but the basic units would be
variations on standard forms that the manufacturer produces.

For the type of unit shown here, typical variables would be the actual
width of the T flange and the depth of the T stem. Special steel elements
would be cast into the units to permit attachment to supports, attachment
of adjacent units to each other, and possibly attachment of some architectural elements for the windows, roof edge details, and so on.

The roof units deliver a considerable concentrated load to the tops of the walls. A primary design investigation for the walls would be the effect of this loading and the required thickness of the wall. An investigation of this type is done as the Example problem in Section 12.2. In that problem, it was found that a minimum thickness for the bearing condition was 6 in. However, the wall must also sustain lateral bending due to wind pressure or seismic effects. In addition, it must be poured in forms, and the 6-in. thickness is probably not feasible for the 12-ft height. A maximum height-to-thickness ratio of 15 is usually recommended, so a thickness of 10 in. would be more practical.

The combination of the very heavy roof structure and the heavy concrete walls will result in a considerable load to the wall footings. Settlement of these footings should be dealt with very conservatively because the heavy exterior structure will tend to settle more than the lighter interior construction.

The flange portions of the roof T units can be attached to form a horizontal diaphragm. This is done by welding of steel elements cast into the T flange edges. A leveling concrete fill is usually used on top of the structure, because the dead load deflection of the individual T units can be controlled only very approximately. However, for a roof, this may be a very lightweight, insulative fill, rather than one with significant strength.

The exterior concrete walls, with their turned corners in plan, constitute a very effective, lateral bracing system in both directions. Overturn is surely not a problem with the dead weight of the construction. The major concern here is for transfer of the lateral force from the roof to the walls through the seat connections for the T units.

For the exposed walls, an architectural design concern involves the visibility of forming elements. Walls will most likely be formed with panels using 4 ft by 8 ft sheets of plywood, and this unit is hard to conceal on the surface. It is better to exaggerate the joints between panels and make them a part of some modular pattern on the wall surface. The forms for opposite wall faces are ordinarily tied with metal devices extending through the walls. When the forms are removed, the ends of these ties will be visible on the wall surface. There is no real possibility of hiding them on the exposed concrete surface, so they should be made a part of the wall surface pattern as well. Figure 16.22 shows a partial elevation of the wall with a layout of the form joints and ties that is coordinated with the module of the roof units.

With the absence of a suspended ceiling construction, it is necessary to attach various elements directly to, or suspend them from, the T units,

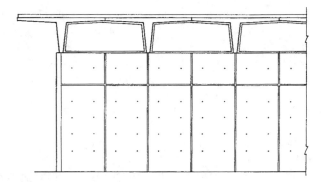

Figure 16.22 Building Three, Alternative One: exterior wall surface developed to express the forming of the wall joints and form ties.

for lighting, HVAC services, fire sprinklers, signs, and so on. Provision for general attachment may be made by casting some elements into the underside of the T units.

The following sections illustrate some variations on the structure for this building that maintain the basic form of the building plan.

16.8   BUILDING THREE: ALTERNATIVE STRUCTURE TWO

Figure 16.23 shows an alternative structure for Building Three, using a concrete waffle slab system for the roof. Two interior columns are used to define a six-bay system for the waffle, with edges supported by a series of columns in the exterior walls. The exterior walls are formed with precast, tilt-up units, with the columns cast in place to connect the precast units.

The waffle system is described in detail in Section 9.4. The system shown here is one commonly produced with square, metal "pan"-type forms. Pans are 30 in. square in plan, and when spaced on 3-ft centers, produce 6-in.-wide ribs between the voided spaces in the waffle. Wider ribs can be produced by simply spacing the pans farther apart. Other pan sizes can also be used for special situations. As shown in the construction section in Figure 16.23, a wider rib is provided over the exterior wall, parallel to the roof edge; this is to facilitate the joint with the columns at this location.

Referring to the plan in Figure 16.23, it may be observed that the waffle system is supported on the building perimeter by columns at 16-ft centers. A cantilevered overhang is developed with one unit of the waffle

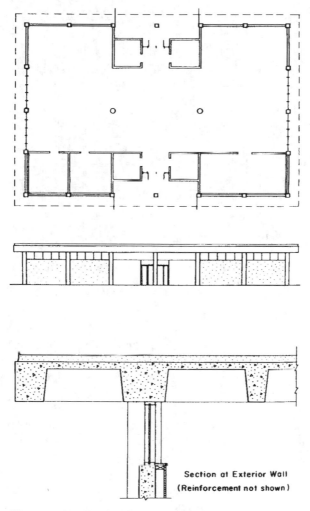

Figure 16.23 Building Three, Alternative Two: general form.

outside the columns, as shown in the construction section. The general pattern of the waffle system is illustrated in the reflected plan in Figure 16.24 (like looking in a mirror on the floor).

The interior columns serve to define a series of 32-ft square bays for the two-way spanning waffle system. The solid portions around the interior columns and corresponding exterior columns serve to strengthen the

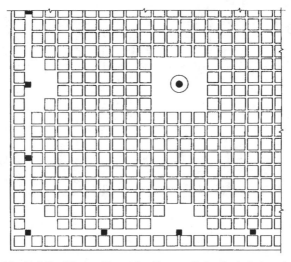

Figure 16.24   Building Three, Alternative Two: partial reflected plan showing the layout of the waffle system.

system for the concentration of shear and bending in the spanning system. (See the general discussion of two-way spanning systems in Section 9.3.) Some alternatives for development of the structure at the interior column are shown in Figure 16.25 as follows:

1. Figure 16.25*a* indicates a shift of the waffle layout to have a void/coffer rather than a rib on the column line, permitting a smaller solid portion at the column than that shown in Figure 16.24. This reduces the width of the solid portion below that ordinarily recommended, so the shear and bending would need to be investigated to demonstrate that this solution is adequate.

2. Figure 16.25*b* indicates a retaining of the layout in Figure 16.24, except that 20-in. square pans are used in the corners of the solid portion to reduce its mass slightly and ease the visual transition from the waffle to the solid portion. This tends to pull the waffle pattern into the solid portion.

3. In a similar fashion, Figure 16.25*c* shows the use of 20-in. square pans to extend the solid portion shown in Figure 16.25*a*. The 20-in. pans are the other standard-size pan, ordinarily used with 4-in.-wide

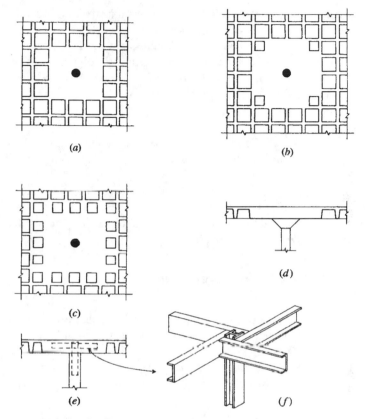

Figure 16.25 Building Three, Alternative Two: alternative details at the interior column.

ribs to form a 24-in. modular rib pattern. Pans are also available in other special dimensions, to facilitate special edge conditions (for example, 10 by 10, 10 by 20, 10 by 30, 15 by 15, and so on).

4. Another technique is to strengthen the resistance to shear by widening the tops of the columns, also slightly reducing the span in the process. This is a trick developed by ancient builders with stone to allow the tension-weak stone lintels to work with wider-spaced columns. Figure 16.25d shows the use of a spread-formed capital on the column. This is commonly done with flat slab construction (see Section 9.3).

5. Another way of achieving the column capital effect, where the capital as shown in Figure 16.25*d* may not be desirable, is to cast a steel element into the solid portion, as shown in Figure 16.25*e*. A possible form of such an element is shown in Figure 16.25*f*, produced by welding steel rolled sections. Although shown embedded in a concrete column here, this device could also be used with a steel column.

These and other variations are possible considerations for either improving the structural performance of the joint or simply manipulating its visible form.

The exterior columns would be detailed and constructed as ordinary reinforced columns, as discussed in Chapter 10. If the infill exterior wall is nonstructural in nature, these columns could be made to constitute a rigid frame with the widened waffle ribs.

16.9 BUILDING THREE: ALTERNATIVE STRUCTURE THREE

Figure 16.26 shows another possibility for the construction of Building Three, replacing the cast-in-place concrete walls shown in Figure 16.21 with walls of reinforced concrete masonry. In this solution, the concentrated loads from the T units could be resisted by the internal reinforcement of the walls or by developing pilaster columns, as shown in Figure 16.26*b*.

The masonry walls will function just as the concrete walls and will be designed for the gravity and lateral loads in a similar manner. A concern for this structure is the development of a modular, dimensional system for plan and vertical layouts, so that full CMUs can be used. Common vertical units (individual block heights) are 4, 6, and 8 in. The usual unit in the direction of the wall plane in plan (block length) is 8 in., with the standard block being 16 in. long. The single-unit-thick wall is typically developed with 6-, 8-, 10-, or 12-in.-wide blocks.

More possibilities for details of special elements of this structure are illustrated in the development of the masonry structure for Building Four, as described in Section 16.11. The masonry walls and columns shown here could also be used with the waffle structure shown in Section 16.8. In all, there is a considerable number of alternatives for this building if the whole range of possible mixtures of sitecast concrete, precast concrete, and masonry are considered.

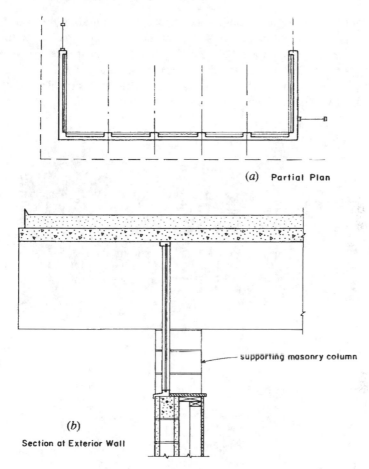

(a) Partial Plan

supporting masonry column

(b)
Section at Exterior Wall

Figure 16.26 Building Three, Alternative Three: partial plan and construction details.

16.10 BUILDING FOUR: GENERAL CONSIDERATIONS

This section presents discussion of the development of construction for a three-story office building. The construction illustrated shows the use of mixed elements of wood, masonry, and concrete.

The architectural predecessor for this building is a form of construction described as *mill construction,* so-called because it was first developed for use in industrial buildings in the early years of the Industrial

Revolution. As originally developed, most of these buildings used heavy masonry walls and timber roof structures. For multistory buildings, timber was also used for the interior columns and floor structures. Some examples of this form of construction are shown in Figures 16.27 and 16.28.

Figure 16.27 shows the general form and some details for a multistory building with brick masonry walls and a roof and upper floors developed

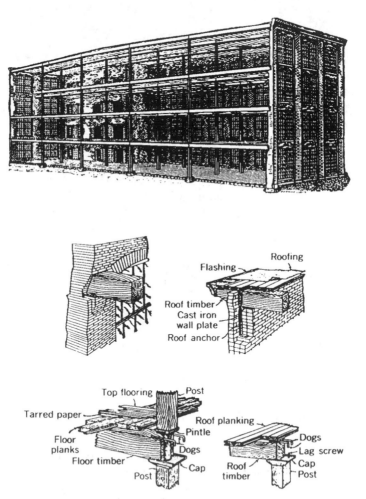

Figure 16.27   Example of mill construction for a multistory building. Reproduced from *Construction Revisited* with permission of the publishers, John Wiley & Sons.

with post-and-beam construction in timber. Another possibility for the multistory building was the use of longspan trusses for the roof, providing a large, open floor on the top story of the building. Figure 16.28 shows examples of truss systems used for single-story buildings.

Figure 16.29 shows some details for Building Four, indicating the use of masonry walls; interior concrete columns; upper floors of reinforced

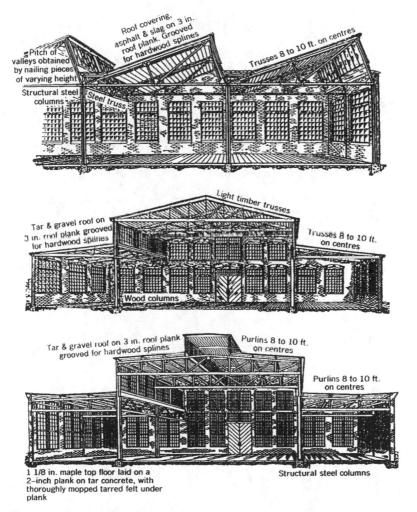

Figure 16.28 Examples of mill construction for a one-story building. Reproduced from *Construction Revisited* with permission of the publisher, John Wiley & Sons.

concrete, slab and beam form; and a timber truss roof system spanning
the full width of the building. The upper floors are open at the center of
the building, providing a clear height from the first floor to the skylight
at the ridge of the trussed roof.

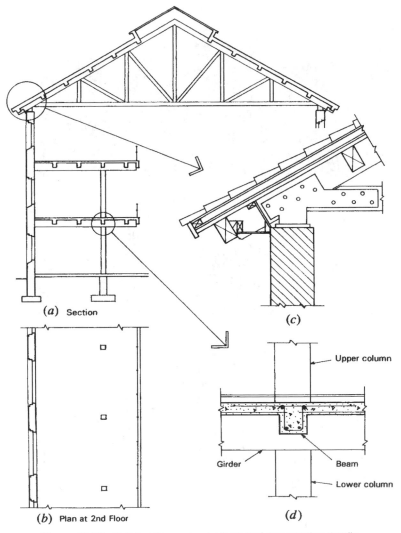

Figure 16.29 Building Four: general form and construction details.

The following sections provide discussions of the design of the masonry walls, the upper concrete floors, and the development of the lateral loads on the structure.

16.11   BUILDING FOUR: CONCRETE AND MASONRY STRUCTURE

The horizontal floor structures for the upper levels are developed as sitecast reinforced concrete. Interior columns are also sitecast concrete. The exterior wall structure is a composite of structural masonry, developed with CMUs, and integral elements of cast concrete.

The Concrete Floor Structure

A partial plan of the concrete framing for one of the upper floors in Building Four is shown in Figure 16.30.

A series of beams, 16 ft on center, is supported by pilasters in the exterior masonry walls and extends through matching interior concrete columns to support the 8-ft cantilever portion of the floor. These main beams support a series of closer-spaced beams, which in turn supports the concrete deck.

The spans for this system are quite modest, so the slab and beams will be of minimum dimension. For fire resistance requirements, the slab is likely to be not less than 4 in. thick, which is more than adequate for the

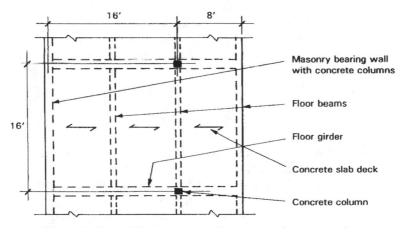

Figure 16.30   Building Four: partial framing plan for an upper floor.

gravity load design for the clear span of less than 8 ft. The computations for this basic system are not shown here because the general form and design process for the slab-and-beam sitecast system is illustrated elsewhere (see discussion in Sections 10.1 and 10.2, and design example for Building Five in Section 16.16).

The pilasters in the exterior wall that provide support for the ends of the main floor beams are actually concrete columns developed within the CMU walls. These and other details for the intersection of the concrete floor and exterior walls are treated in the discussion of the masonry walls that follows.

A special problem is that of developing the horizontal bracing system for lateral loads at the upper floor levels. This is discussed in Section 16.12.

The Composite Masonry and Concrete Walls

A partial elevation of one of the exterior walls is shown in Figure 16.31. The three-story-high pilasters are expressed as vertical ribs at 16-ft centers on the building exterior. A wide strip at the top of the masonry wall is also thickened to match the pilasters.

Plan sections of the exterior walls that illustrate the formation of the construction are shown in Figure 16.32. The two upper sections in the figure show the layouts for alternating courses of the masonry units as they occur at the level of the windows. The lower section in Figure 16.32 shows one of the courses of masonry units at the continuous solid wall in the spandrel area between windows.

Although technically formed within the CMU construction, the pilasters actually consist of reinforced concrete columns of significant dimension, probably about 13 in. square within the nominal size, 16-in. square masonry units. These columns may be visualized to be parts of the interior concrete structure; indeed, they may be cast together with that system.

Extension of the horizontal reinforcement in the wall through the pilasters will make the concrete pilasters integral parts of the wall construction. The relatively closely spaced pilasters will, therefore, provide considerable stiffening for the walls for both gravity loads and shear wall actions.

Vertical section details of the masonry walls are shown in Figure 16.33. Although the edge of the concrete floor structure appears to rest on the projected edge of the masonry, the edge beam is actually supported by the main floor beams, which in turn are supported by the concrete col-

Figure 16.31   Building Four: elevation of the exterior wall with concrete masonry unit (CMU) construction.

umns inside the pilasters. Thus, the entire floor structure transfers vertical gravity load only to the pilasters.

Horizontal beams in the walls are developed as thickened sections. These beams form a frame with the pilasters for the development of shear wall actions and the transfer of the loads of the wall weight to the columns. In the event that the trusses cannot be located over the pilasters, the beam at the top of the wall could be developed to span between pilasters to support the truss loads.

The infill walls of 8-inch CMUs are essentially nonstructural in nature for gravity loads, although it is probably advisable to develop them with the usual form of reinforced masonry construction (both vertical and horizontal reinforcement in grout-filled cores). For lateral loads in the plane of the walls, these infill panels will so stiffen the wall that they will truly

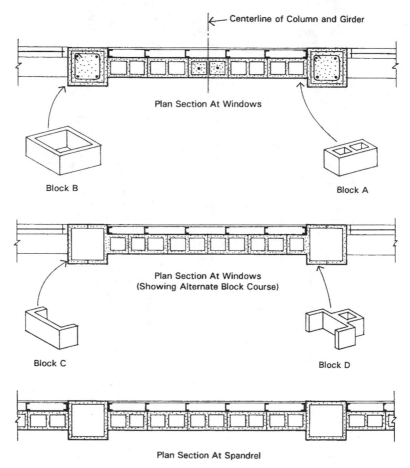

Figure 16.32   Building Four: plan section details of the exterior CMU walls.

interact with the pilasters and horizontal beams. The net effect will be a shear-wall-stiffened frame. For the size of this building and the amount of solid wall surface, as shown in Figure 16.31, this shear wall function can probably be provided with minimum CMU construction.

The Foundations

The details in Figure 16.33 show a concrete floor slab on grade for the first floor and a grade wall developed by extension of the CMU con-

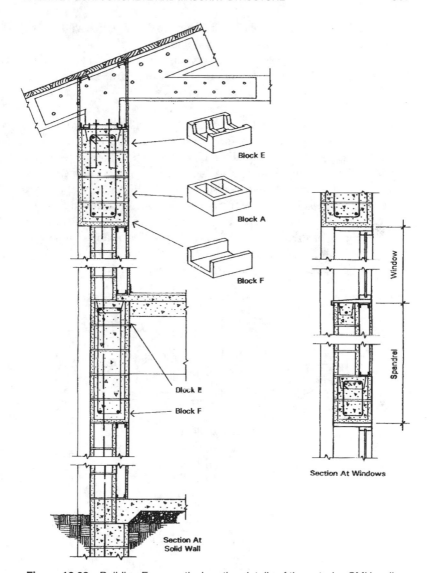

Figure 16.33 Building Four: vertical section details of the exterior CMU walls.

struction. This detail is also shown in Figure 16.34, together with a complete section of the grade wall and wall footing.

Although the pilasters at 16-ft centers carry concentrated loads to this footing, it is possible to use the reinforced masonry grade wall as a distributing stiffener and to design the footing as a single strip for a continuous soil pressure. This is the solution shown here.

This is a quite heavy wall. The concrete floors, heavy roof, and masonry walls will generate a considerable dead load. The relatively small upper floor areas and the roof will develop only a minor live load. Design here, therefore, should be for mostly the long-term effects of the permanent dead load. A critical consideration should be for settlement because of the high dead load and the sensitivity of the stiff construction to deformations. A highly conservative value for design soil pressure is advised.

An estimate of the total load for the wall footing is approximately 8 kips per ft. With a soil pressure of 2 ksf, this requires about a 4-ft-6-in.-wide footing. Referring to Table 13.4, a possible selection is:

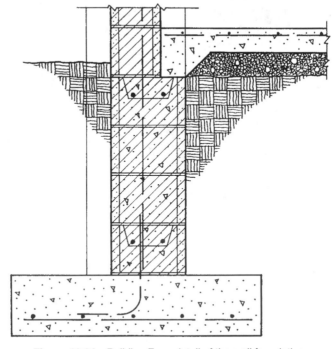

Figure 16.34   Building Four: detail of the wall foundation.

- 4-ft-by-6-in.-wide, 11-in.-thick concrete section
- Three No. 5 bars in the long direction
- No. 5 bars at 11-in. centers in the short direction

A column footing must be provided for the interior column. Although a pedestal is shown for this in the section in Figure 16.29, this would be used with only a wood or steel column. In this example, the concrete column may be simply extended down as far as required to the top of the footing. This is a modestly loaded column and footing, so the column can probably be as small as is feasible for reinforced concrete (10 to 12 in. square or round). The total load for the lower-story column and the footing here is about 60 kips. For the same design pressure of 2 ksf as used for the wall footing, Table 13.6 shows a possibility for the following:

- 6-ft square, 14-in.-thick footing
- Six No. 6 bars each way

16.12 BUILDING FOUR: DESIGN FOR LATERAL LOADS

For this building, design for lateral forces of either wind or earthquakes involves the following three major concerns:

1. Design of the masonry walls as shear walls, with combined lateral and gravity load functions
2. Design of elements of the wood roof and concrete floor structures to serve necessary diaphragm, collector, tie, and other lateral load resisting functions
3. Design of connections between the roof and floor framing and the masonry walls for the necessary transfer of forces due to lateral loads, in combination with any forces due to gravity loads

A complete illustration of the design of this structure for lateral loads (and lateral load + gravity load combinations) might entail a presentation the size of this entire book. Much of the work is illustrated in other examples here: determination of loads, investigation for forces in diaphragms, analysis for anchorage requirements, and so on.

What is not developed elsewhere in this book has mostly to do with the problems of the hollowed-out building, as shown in Figure 16.29. The discussion here is, therefore, limited to this concern.

The floor structure at ground level and the roof structures are generally continuous surfaces. The slope of the roof makes it questionable to consider the roof deck as an actual horizontal diaphragm, but this is generally the assumed case as long as the slope does not much exceed the slope shown here. However, the lateral stability of the tops of the exterior walls will be much improved if horizontal trussing is provided at the level of the bottom chords of the roof trusses.

The floor structures at the second- and third-floor levels consist of donut-shaped surfaces. A small hole may be incorporated in either a horizontal or vertical diaphragm, but as the size of the hole increases, some different functioning of the structure must be considered. There is no clear line here, although the extreme cases are reasonably definite (see Figure 16.35).

One approach for the upper floors here is the possibility of allowing the floor at a single level to consist of a connected set of smaller diaphragms within the total surface (called *subdiaphragms*). If these subdiaphragms are individually capable of their assigned tasks, the system is probably reasonable. With the size and proportions of the surfaces here, this is probably a reasonable assumption. The loads here will also not be excessive in magnitude for the three-story building.

Although not often feasible with a wood structure, the possibility exists for turning the entire floor structure into a single rigid frame, as is commonly done with vertical planar bents of columns and beams (see discussion in Section 16.16). This may be achieved with the beam system alone or with consideration of the additional effect of the decking.

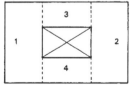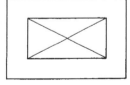

Small Opening:	Large Hole:	Very Large Hole:
functioning of whole diaphragm mainly unaffected; reinforce edges and corners of hole	diaphragm reduced to parts (subdiaphragms) that may work as a connected set	not a diaphragm, can function only as a very stiff rigid frame

Figure 16.35 Range of effects of a hole in a horizontal diaphragm: (*a*) small hole with negligible effect on the diaphragm action, (*b*) large hole, possibly indicating the need for subdiaphragms, and (*c*) hole so large that the structure is incapable of functioning as a diaphragm and may need to function as a horizontal rigid frame.

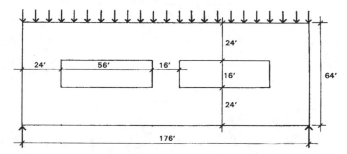

Figure 16.36 Building Four: diaphragm action of the floor structure.

Figure 16.36 shows a plan of the typical upper floor, indicating its use as a horizontal spanning unit from one end of the building to the other. An approach here is to consider the concrete slab as taking the beam shear on its net width at all points along the beam.

A possible concern at the perimeter of the floor is the ability of the edge beam to function as a compression or tension member as the diaphragm chord. For this purpose, it should have some continuous reinforcement, and for the compression, it should have full loop ties at relatively close spacing.

A possible additional concern is for the functioning of some subunits of the floor structure. The 56-ft-long narrowed floor portion here, for example, should be able to function as a subdiaphragm, spanning on its own for the 56-ft distance to carry the edge loading.

As mentioned previously, there is no horizontal diaphragm at the bottom of the trusses. This function must be provided by the sloping roof deck or by some bracing in a horizontal plane at the level of the bottom of the trusses. In this example, another possibility is to develop the top story of the pilasters as cantilevers, taking the horizontal force from the trusses at the tops of the columns. This is only about an 8-ft cantilever in this example, so it may be a reasonable solution.

Because of its weight, this structure may be in more serious trouble if it needs to sustain seismic forces for a high-risk zone. The same systems may be used for bracing, but forces and stresses may well be much higher.

16.13 BUILDING FOUR: ALL-CONCRETE STRUCTURE

Many forms of construction can be used to produce a building with the appearance shown in Figure 16.31. The masonry exterior can be developed as an applied veneer over a variety of structures. One such variation

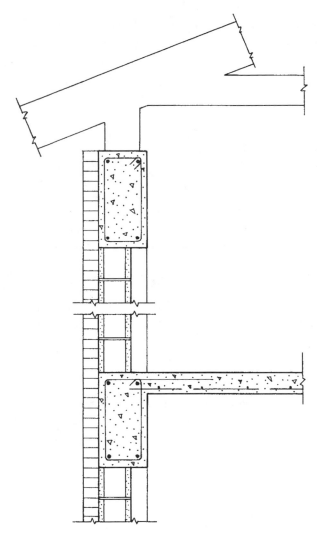

Figure 16.37 Building Four: construction details for the all-concrete structure with masonry veneer walls.

is shown in the wall section details in Figure 16.37. This is a modification of the wall section shown in Figure 16.33.

In Figure 16.37, the basic building structure—from the underside of the trusses to the bottom of the footings—is developed fully with reinforced concrete. Actually, the modification consists essentially of differences in the exterior wall structure because the interior concrete columns and the framing for the upper floors would remain as shown previously.

The masonry veneer shown in Figure 16.37 could as well be installed over a steel frame, and frequently is. Alternately, a variety of exterior curtain wall systems could be placed on the concrete frame shown here. Some variations on this are illustrated for Building Five.

Masonry surfaces for buildings are popular—or at least the *appearance* of masonry. As discussed in Section 12.6, the appearance of masonry can be achieved in a variety of ways that vary in terms of their identity as real masonry. The construction shown in Figures 16.32 and 16.33 is quite close to that of real structural masonry, modified only by the reinforced concrete pilasters. The construction shown in Figure 16.37 uses real masonry for the building exterior surface; it is simply not in the classification of structural masonry.

A variation on the opposite end of the spectrum of "reality" is that shown in Figure 16.38. This construction uses a form described as exterior fiber-reinforced insulation (EFI), which is quite popular and truly energy-smart because it keeps the bulk of the wall construction inside as a thermal inertial mass. This is the ultimate Disneyland exterior, which can assume just about any desired appearance. In fact, the elevation shown in Figure 16.31 can easily be achieved by simply adhering thin, brick tiles to the outside of this construction.

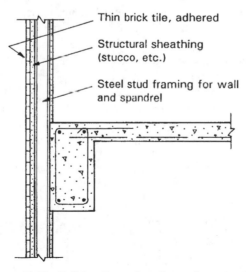

Thin brick tile, adhered

Structural sheathing (stucco, etc.)

Steel stud framing for wall and spandrel

Figure 16.38   Building Four: alternative wall construction.

364 BUILDING STRUCTURES: DESIGN EXAMPLES

16.14 BUILDING FIVE: GENERAL CONSIDERATIONS

Building Five is a three-story office building, planned for speculative rental. As with preceding examples, there are many alternatives for the construction of such a building, although in any given place at any given time, the basic construction will most likely vary little from a limited set of choices. Shown here is some of the work for the design of two alternatives for the construction: one using structural masonry walls and the other using a complete sitecast concrete structure.

General Considerations for the Building Design

Figure 16.39 presents a plan of the upper floor and a full building section for Building Five. A general requirement for office buildings is the provision of a significant amount of exterior window surface and the avoidance of long expanses of solid wall surface. This building is assumed to be free-standing on its site, so this applies to all sides of the building exterior.

Most designers and their clients would also prefer that the space available for rental be as free as possible of interior permanent construction, thus permitting maximum flexibility in rearrangement of the interior for future tenants. For the structure, this translates into a desire to eliminate permanent structural elements (columns and structural walls) from the rental spaces. When possible, permanent building elements, such as restrooms, stairs, elevators, and vertical risers for services, are grouped near the building center, where rental space is less desirable.

The exterior walls of the building are permanent, but do not intrude on the rental floor space. Except in relation to planning of offices to match the window modules, the exterior walls are of prime concern only to the architect and the structural designer. There is considerably more freedom here for location and shaping of structural elements, both for the vertical structure and the horizontal structure—that is, both for columns and spandrel beams.

For the work illustrated here, the following criteria are assumed:

- Design Codes: 1997 UBC (Reference 3) and ASCE 2002 (Reference 2)
- Live Loads:
  - Roof: Minimum code required, see Section 15.6
  - Floors: See Table 15.2
  - Office areas: 50 psf
  - Corridors and lobby: 100 psf

- Partitions: 20 psf
- Wind: Map speed of 90 MPH, Exposure B
- Assumed construction loads:
  - Floor finish: 5 psf
  - Ceilings, lights, and ducts: 15 psf
- Walls (average surface weight):
  - Interior, permanent: 10 psf
  - Exterior curtain wall: 15 psf

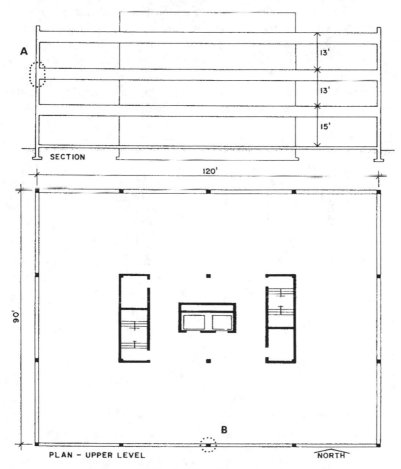

Figure 16.39 Building Five: general form.

Structural Alternatives

The plan as shown, with 30-ft square bays and an open interior, is an ideal arrangement for a beam-and-column frame system in either steel or concrete. Other types of structural systems may be made more effective if some modifications of the basic building plan are made. Such changes may affect the planning of the building core (stairs, elevators, etc.), the plan dimensions for column locations, the form of the exterior walls, and the vertical distances between levels (stories) of the building.

The general form and basic type of the structural system must relate to the resistance of both gravity and lateral forces. Considerations for gravity affect the spans of the horizontal structure and the arrangement of vertical support elements. Vertical columns and bearing walls should be stacked, thus requiring coordination of the plans for the various levels of the building.

The most common options for the lateral bracing system for a multistory building are the following (see Figure 16.40):

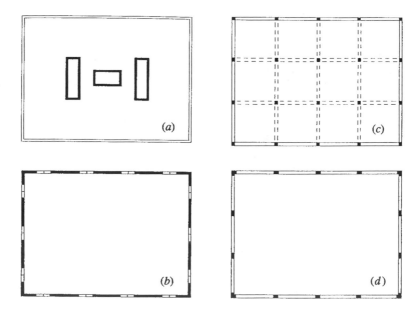

Figure 16.40 Options for the vertical elements of the lateral bracing system: (a) braced core with shear walls or trussed bents, (b) braced perimeter with shear walls or trussed bents, (c) fully developed, three-dimensional, rigid frame, and (d) perimeter rigid frames.

1. *Core Shear Wall System* (Figure 16.40*a*). This consists of using solid walls to produce a very rigid central core tower. The rest of the structure then leans on this rigid interior portion, attached through the horizontal structure at each level. This leaves the perimeter structure and most of the horizontal structure to be designed primarily for gravity loads. However, the collecting of the lateral forces—for example, from wind pressures on the building exterior surface—will involve some other elements of the structure.

2. *Truss-Braced Core.* This is similar to the shear wall core in terms of development of the rest of the structure.

3. *Perimeter Shear Walls* (Figure 16.40*b*). This makes the building into a tubelike structure. The perimeter bracing walls may be developed as continuous walls with openings or as a series of individual wall piers.

4. *Mixed Perimeter and Interior Shear Walls.* For buildings with long, narrow plans or very complex forms, a combination of interior and perimeter shear walls may be used. This is simply a combination of the schemes in Figures 16.40 *a* and *b*.

5. *Rigid Frame System* (Figure 16.40*c*). This is produced by using vertical planes of rigidly attached columns and beams to produce a series of two dimensional rigid frames. A full rigid frame system uses all of the columns and column-line beams. However, selected planar frames can also be used simply by increasing their relative stiffness. Distribution of the lateral forces will occur in proportion to the stiffness of the individual frames. Thus, very rigid frames will attract the greater loads.

6. *Perimeter Rigid Frame System* (Figure 16.40*d*). This consists of using the exterior columns and spandrel beams as the primary bracing system. This is usually achieved by stiffening the exterior column/beam bents. It is similar in character to the perimeter shear wall system.

In the right circumstances, any of these systems may be acceptable. Each option has advantages and disadvantages, from both architectural and structural design points of view. The core-braced systems were popular in the past, especially for tall buildings, for which wind load is usually the critical concern. For the architect, a major advantage of the core-braced system is the relative freedom of planning of the building exterior form. However, all the other options have the potential for producing structures that are stiffer and have more resistance to torsion on the building. The latter is most significant where earthquake forces are critical.

16.15   BUILDING FIVE: MASONRY AND FRAME STRUCTURE

A structural framing plan for one of the upper floors in Building Five is shown in Figure 16.41. The plan indicates the use of bearing walls as the major supports for the floor framing. The walls will also constitute the lateral bracing system, with a combination of perimeter and core-braced features.

For the office building, there are many concerns for the incorporation of the elements of the systems for wiring, piping, ducting, fire control, and lighting. Also of concern are the developments of floor finishes and ceiling construction, as well as the necessary fire rating for the structure and the overall construction. There are many options for the floor structure in terms of both its structural merits and the ability to accommodate the rest of the building construction. For this example, we will assume that the various considerations can be met by a horizontal system consisting of a plywood floor deck; light, nailable prefabricated joists or trusses and steel beams, as shown in Figure 16.42.

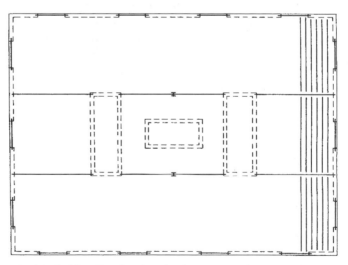

Figure 16.41   Building Five: framing plan for the upper floors with the masonry wall structure.

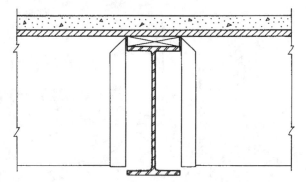

Figure 16.42 Building Five: detail of the upper floor construction with the masonry wall structure.

Figure 16.43 shows the general construction of the exterior walls, indicating the use of reinforced CMU construction for the wall structure. A multilayered EFI is used on the outside and furring strips with gypsum drywall on the inside. The remainder of the discussion for this example deals with the design of the structural masonry walls.

Design for Gravity Forces

There are two general concerns for the walls for gravity loading: first, for the compressive stress in the wall, and second, for the effects of concentrated loads from supported beams. For compression, the greatest load will be in the first story, and this condition will be the basis for choice of the wall thickness and the rated class of the masonry. From this maximum condition, the wall may be graded down in terms of compression capacity in upper stories. Graduation may be achieved by changing the thickness or the class of the masonry or by modifiying of the concrete fill and reinforcement.

A minimum capacity is established by code limitations for the minimal construction, as described in Sections 2.9 and 10.14. With standard 16-in.-long, two-void blocks, voids are 8 in. on center. This becomes the basic unit for horizontal planning of the walls. Codes require a minimum of concrete-filled voids with reinforcement every 4 ft, both horizontally and vertically. From that minimum, additional voids may be filled, up to a maximum condition with all voids filled. The latter is usually required

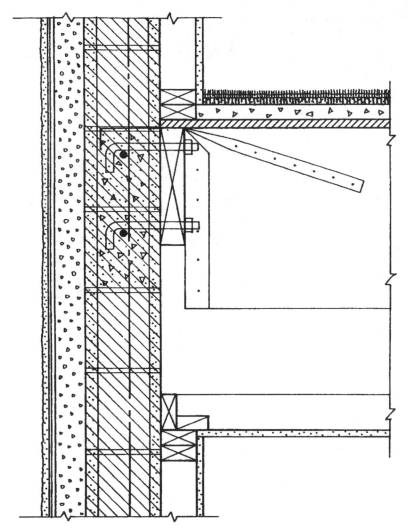

Figure 16.43 Building Five: optional details for the exterior masonry walls.

for heavily loaded shear walls. It is also possible to increase the reinforcement from a minimum amount.

As the framing plan in Figure 16.41 shows, the interior steel beams and the lintels are supported at ends or corners of the bearing walls. These locations will automatically have filled and reinforced voids with some column compression capacity. However, it is also possible to provide enlarged elements (pilasters) at these locations, as illustrated for the walls in Building Four (see Figures 16.31 and 16.32).

Although design for gravity must be individually considered, the walls must also sustain the combination of gravity plus lateral loads. Development of the wall system for lateral loads is considered next.

Design for Lateral Forces

For this example, the walls will be designed for the effects of wind loads, as described in Section 15.7. For the total wind on the building, we will assume a base pressure of 15 psf, adjusted for height as described in ASCE 2002 (Reference 2). The design pressures and their zones of application are shown in Figure 16.44.

For investigation of the lateral bracing system, the wind pressures on the surface of the building are distributed as edge loadings to the roof and floor horizontal diaphragms. The total loads to the diaphragms are shown as the forces H_1, H_2, and H_3 in Figure 16.44.

The plan in Figure 16.45a illustrates an assumed distribution of the total wind load to the groups of end walls and core walls on the basis of peripheral distribution by the wood deck. The zones for this distribution are shown in Figure 16.45b. This loading will put a slightly higher load on the core walls.

Distribution is determined in terms of a relative stiffness based on the plan lengths of the walls (shown in Figure 16.45a) and their height-to-width ratios, as indicated in Figure 16.45c. Note that although the core walls actually have tubelike forms, the walls used in this distribution are only those parallel to the wind direction.

Although the distribution of shared loads to the walls is usually done on the basis of their relative stiffness, an average shear stress may be obtained by simply dividing the total shared load by the sum of the wall lengths. On this basis, the average shear stress is

$$v = \frac{70,150}{260} = 270 \text{ lb/ft of wall length } [3.94 \text{ kN/m}]$$

This is a low stress for a reinforced CMU wall. Even though the actual stress will be somewhat higher in the stiffer walls, this computation indicates that there is considerable redundant strength in the construction for resistance to wind loads.

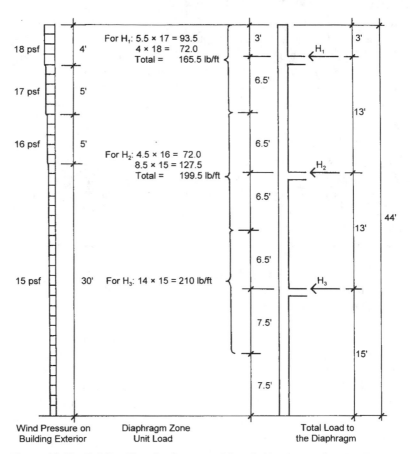

Figure 16.44 Building Five: development of the wind loads transferred to the roof and floor deck diaphragms. Design wind pressures on the building exterior from ASCE 2002 (Reference 2), assuming a base wind pressure of 15 psf. Diaphragm zones are defined by column midheight points. Total loads on the diaphragm at each level are found by multiplying the diaphragm zone unit load per foot by the width of the building.

If the individual wall elements shown in Figure 16.45a act as individual piers, they must be considered individually for overturn effects. For the building height shown, the 10-ft walls will be quite slender if intended to work for all three stories. However, they are really not needed for the lower magnitude of lateral force in the upper floors, so they could be considered as acting only for the first-story shear force.

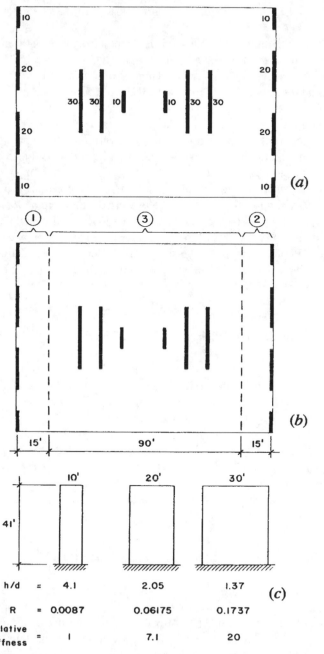

Figure 16.45 Building Five: considerations for the distribution of lateral loads to the north-south shear walls.

If the exterior walls are developed as continuous walls with small openings for the windows, their lengths on the plan in Figure 16.45a should be shown as 92 ft. In this case, the lateral stiffness of the core walls would become insignificant, and the end walls would be assumed to take the entire lateral load.

With the building almost twice as wide as it is tall in the narrow direction, there is no concern for a general overturn of the whole building. This is a major concern for very tall, towerlike buildings, but not usually critical for low-rise structures.

A truly critical issue for this construction is the connection of the horizontal framing to the walls. There is a three-way force action here, involving vertical gravity force, lateral shear in the wall plane, and a pullaway effect due to outward force on the wall (wind suction or earthquake effect). Participation of all the structural elements in these actions (wall, anchor bolts, deck, joists, beams, and ledgers) must be studied and adequately designed for these effects.

16.16   BUILDING FIVE: THE CONCRETE STRUCTURE

A structural framing plan for the upper floors in Building Five is presented in Figure 16.46, showing the use of a sitecast concrete, slab and beam system. Support for the spanning structure is provided by concrete columns. The system for lateral bracing is that shown in Figure 16.40d, which uses the exterior columns and spandrel beams as rigid frame bents at the building perimeter. This is a highly indeterminate structure for both gravity and lateral loads, and its precise engineering design would undoubtedly be done with a computer-aided design process. The presentation here treats the major issues and illustrates an approximate design, using highly simplified methods.

Design of the Slab-and-Beam Floor Structure

For the floor structure, use $f'_c = 3$ ksi [20.7 MPa] and $f_y = 40$ ksi [276 MPa]. As shown in Figure 16.46, the floor framing system consists of a series of parallel beams at 10-ft centers that support a continuous, one-way spanning slab, and are in turn supported by column-line girders or supported directly by columns. Although special beams are required for the core framing, the system is made up largely of repeated elements. The

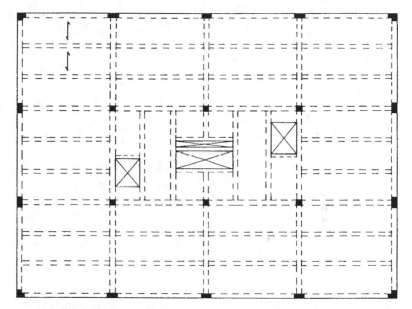

Figure 16.46 Building Five: framing plan for the concrete structure for the upper floor.

discussion here will focus on three of these elements: the continuous slab, the four-span interior beam, and the three-span spandrel girder.

Using the approximation method described in Section 9.1, Figure 16.47 shows the critical conditions for the slab, beam, and girder. Use of these coefficients is reasonable for the slab and beam, which support uniformly distributed loads. For the girder, however, the presence of major concentrated loads makes the use of the coefficients somewhat questionable. An adjusted method is, thus, described later for use with the girder. The coefficients shown in Figure 16.47 for the girder are for uniformly distributed loads only (the weight of the girder itself, for example).

Figure 16.48 shows a section of the exterior wall that illustrates the general form of the construction. The exterior columns and the spandrel beams are exposed to view. Use of the full available depth of the spandrel beams results in a much stiffened bent on the building exterior. As will be shown later, this is combined with the use of oblong-shaped columns at the exterior to create perimeter bents that will indeed absorb most of the lateral force on the structure.

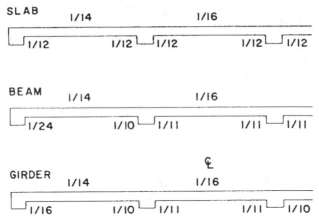

Figure 16.47 Approximate design factors for the slab and beam floor structure.

The design of the continuous slab is presented as the example in Section 9.1. The use of the 5-in. slab is based on assumed minimum requirements for fire protection. If a thinner slab is possible, the 9-ft clear span will not require this thickness, based on limiting bending or shear conditions or recommendations for deflection control. If the 5-in. slab is used, however, the result will be a slab with a low percentage of steel bar weight per square foot—a situation usually resulting in lower cost for the structure.

The unit loads used for the slab design are determined as follows:

- Floor live load: 100 psf [4.79 kPa] (at the corridor)
- Floor dead load (see Table 15.1)
- Carpet and pad at 5 psf
- Ceiling, lights, and ducts at 15 psf
- 2-in. lightweight concrete fill at 18 psf
- 5-in.-thick slab at 62 psf
- Total dead load: 100 psf [4.79 kPa]

With the slab determined, it is now possible to consider the design of one of the typical interior beams, loaded by a 10-ft-wide strip of slab, as shown in Figure 16.46. The supports for these beams are 30 ft on center. If the beams and columns are assumed to be a minimum of 12 in. wide, the

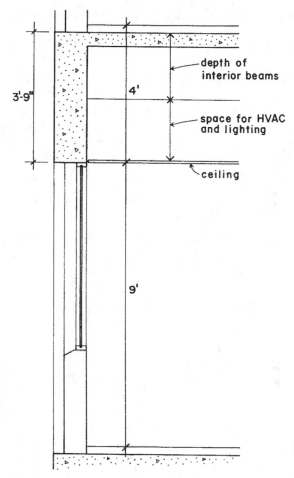

Figure 16.48 Building Five: section at the exterior wall with the concrete structure.

clear span for the beam becomes 29 ft, and its load periphery is 29 × 10 = 290 ft². Using the ASCE 2002 (Reference 2) standard provisions for reduction of live load (see Section 15.6)

$$L = L_0 \left(0.25 + \frac{15}{\sqrt{K_{LL}A_T}}\right) = 100 \left(0.25 + \frac{15}{\sqrt{2 \times 290}}\right) = 87 \text{ psf } [4.17 \text{ kPa}]$$

The beam loading as a per-ft-unit load is determined as follows:

$$\text{Live load} = (87 \text{ psf})(10 \text{ ft}) = 870 \text{ lb/ft [12.69 kN/m]}$$

The dead load without the beam stem extending below the slab is:

$$(100 \text{ psf})(10 \text{ ft}) = 1000 \text{ lb/ft [14.59 kN/m]}$$

Estimating a 12-in.-wide by 20-in.-deep beam stem extending below the bottom of the slab, the additional dead load becomes

$$\frac{12 \times 20}{144} \times 150 \text{ lb/ft}^3 = 250 \text{ lb/ft [3.65 kN/m]}$$

The total dead load for the beam is, thus, $1000 + 250 = 1250$ lb/ft [18.24 kN/m], and the total uniformly distributed factored load for the beam is

$$w_u = 1.2(1250) + 1.6(870) = 1500 + 1392$$
$$= 2892 \text{ lb/ft, or 2.89 kips/ft [42.17 kN/m]}$$

Consider now the four-span continuous beam that is supported by the north-south column-line beams that are referred to as the *girders*. The approximation factors for design moments for this beam are given in Figure 16.47, and a summary of design data is given in Figure 16.49. Note that the design provides for tension reinforcement only, which is based on an assumption that the beam concrete section is adequate to prevent a critical bending compressive stress in the concrete. Using the strength method (see Section 6.3), the basis for this is as follows:

From Figure 16.47, the maximum bending moment in the beam is

$$M_u = \frac{wL^2}{10} = \frac{2.89 \times (29)^2}{10} = 243 \text{ kip-ft [330 kN-m]}$$

$$M_r = \frac{M_u}{\phi} = \frac{243}{0.9} = 270 \text{ kip-ft [366 kN-m]}$$

Then, with factors from Table 6.1 for a balanced section, the required value for bd^2 is determined as

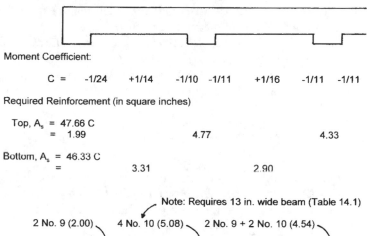

Figure 16.49   Building Five: summary of design for the four-span floor beam.

$$bd^2 = \frac{M}{R} = \frac{270 \times 12}{1.149} = 2820$$

With the unit values as used for M and R, this quantity is in units of in.^{3}. Various combinations of b and d may now be derived from this relationship, as demonstrated in Section 6.3. For this example, assuming a beam width of 12 in. [305 mm]

$$d = \sqrt{\frac{2820}{12}} = 15.3 \text{ in. [389 mm]}$$

With a minimum cover of 1.5 in., No. 3 stirrups, and moderate-size bars for the tension reinforcement, an overall, required beam dimension is obtained by adding approximately 2.5 in. to this derived value for the effective depth. Thus, any dimension selected that is at least 17.8 in. will ensure a lack of critical bending stress in the concrete. In most cases, the specified dimension is rounded off to the nearest full inch, in which case the overall beam height would be specified as 18 in. As discussed in Sec-

tion 6.3, the balanced section is useful only for establishing a tension failure for the beam (yielding of the reinforcement).

Another consideration for choice of the beam depth is deflection control, as discussed in Section 6.3. From Table 6.9, a minimum overall
height of $L/23$ is recommended for the end span of a continuous beam.
This yields a minimum overall height of

$$h = \frac{29 \times 12}{23} = 15 \text{ in. [381 mm]}$$

Pushing these depth limits to their minimum is likely to result in high
shear stress, a high percentage of reinforcement, and possibly some excessive creep deflection. We will, therefore, consider the use of an overall
height of 24 in., resulting in an approximate value of $24 - 2.5 = 21.5$ in.
for the effective depth d. Because this is quite close to the size assumed
for dead load, no adjustment of the previously computed loading for the
beam is made.

For the beams, the flexural reinforcement that is required in the top at
the supports must pass either over or under the bars in the top of the girders. Figure 16.50 shows a section through the beam with an elevation of
the girder in the background. It is assumed that the girder, being much
more heavily loaded, will be deeper than the beams, so the bar intersec-

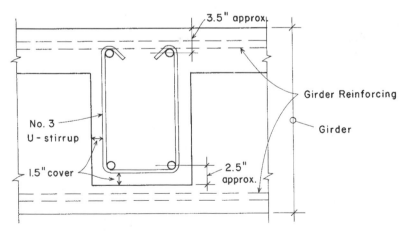

Figure 16.50 Considerations for layout of the reinforcement for the beams and
girders.

tion problem does not exist in the bottoms of the intersecting members. At the top, however, the beam bars are run under the girder bars, favoring the heavily loaded girder. For an approximate consideration, an adjusted dimension of 3.5 to 4 in. should, thus, be subtracted from the overall beam height to obtain an effective depth for design of the beam. For the remainder of the computations, a value of 20 in. is used for the beam effective depth.

The beam cross section must also resist shear, and the beam dimensions should be verified to be adequate for this task before proceeding with design of the flexural reinforcement. Referring to Figure 9.2, the maximum shear force is approximated as 1.15 times the simple span shear of $wL/2$. For the beam, this produces a maximum shear of

$$V_u = 1.15 \frac{wL}{2} = 1.15 \times \frac{2.89 \times 29}{2} = 48.2 \text{ kips } [214 \text{ kN}]$$

As discussed in Section 7.2, this value may be reduced by the shear between the support and the distance of the beam effective depth from the support; thus,

$$\text{Design } V = 48.2 - \left(\frac{20}{12} \times 2.89 \right) = 43.4 \text{ kips } [193 \text{ kN}]$$

and the required maximum shear capacity is

$$V = \frac{43.4}{0.75} = 57.9 \text{ kips } [258 \text{ kN}]$$

With a d of 20 in. and b of 12 in., the critical shear capacity of the concrete alone is

$$V_c = 2\sqrt{f'_c}\, bd = 2\sqrt{3000}\,(12 \times 20) = 26{,}290 \text{ lb, or } 26.3 \text{ kips } [117 \text{ kN}]$$

This leaves a shear force to be developed by the steel equal to

$$V'_s = 57.9 - 26.3 = 31.6 \text{ kips } [141 \text{ kN}]$$

and the closest stirrup spacing at the beam end is

$$s = \frac{A_v f_y d}{V'_s} = \frac{0.22 \times 40 \times 20}{31.6} = 5.6 \text{ in. } [142 \text{ mm}]$$

which is not an unreasonable spacing.

For the approximate design shown in Figure 16.49, the required area of steel at the points of support is determined as follows:

Assume a of 6 in., $jd = d - a/2 = 17$ in. [432 mm]. Then

$$A_s = \frac{M_u}{\phi f_y jd} = \frac{C \times 2.89 \times (29)^2 \times 12}{0.9(40 \times 17)} = 47.66C$$

At midspan points, the positive bending moments will be resisted by the slab and beam acting in T-beam action (see Section 6.5). For this condition, an approximate internal moment arm consists of $d - t/2$, and the required steel areas are approximated as

$$A_s = \frac{M_u}{\phi f_y \left(d - \dfrac{t}{2}\right)} = \frac{C \times 2.89 \times (29)^2 \times 12}{0.9[40 \times (20 - 2.5)]} = 46.33C$$

Inspection of the framing plan in Figure 16.47 reveals that the girders on the north-south column lines carry the ends of the beams as concentrated loads at their third points (10 ft from each support). The spandrel girders at the building ends carry the outer ends of the beams plus their own dead weight. In addition, all the spandrel beams support the weight of the exterior curtain walls. The form of the spandrels and the wall construction is shown in Figure 16.48.

The framing plan also indicates the use of widened columns at the exterior walls. Assuming a minimum width of 2 ft, the clear span of the spandrels, thus, becomes 28 ft. This much-stiffened bent, with very deep spandrel beams and widened columns, is used for lateral bracing, as discussed later in this section.

The spandrel beams carry a combination of uniformly distributed loads (spandrel weight plus wall) and concentrated loads (the beam ends). These loadings are determined as follows:

For reduction of the live load, the portion of floor loading carried is two times one-half the beam load, or approximately the same as one full beam: 290 sq ft. The design live load for the spandrel girders is, thus, reduced by the same amount as it was for the beams. From the beam loading, therefore:

The total factored load from the beam is

$$P = 2.89 \text{ kip/ft} \times 30/2 \text{ ft} = 43.4, \text{ say } 44 \text{ kips } [196 \text{ kN}]$$

The uniformly distributed load is basically all dead load, determined as:

- Spandrel weight: $[(12)(45)/144](150 \text{ pcf}) = 562 \text{ lb/ft } [8.20 \text{ kN/m}]$
- Wall weight: $(25 \text{ psf average})(9 \text{ ft high}) = 225 \text{ lb/ft } [3.28 \text{ kN/m}]$
- Total distributed load: $560 + 225 = 785 \text{ lb}$, say 0.8 kip/ft [11.67 kN/ft]

And the factored load is

$$w = 1.2 \, (0.8) = 0.96 \text{ , say } 1.0 \text{ kip/ft } [14.6 \text{ kN/m}]$$

For the distributed load, approximate design moments may be deter-mined using the moment coefficients, as was done for the slab and beam. Values for this procedure are given in Figure 16.47. Thus,

$$M_u = C(w \times L^2) = C(1.0 \times 28^2) = 784C$$

The ACI Code does not permit use of coefficients for concentrated loads, but, for an approximate design, some adjusted coefficients may be derived from tabulated loadings for beams with third-point load place-ment. Using these coefficients, the moments are

$$M_u = C(P \times L) = C(44 \times 28) = 1232C$$

Figure 16.51 presents a summary of the approximation of moments for the spandrel girder. This is, of course, only the gravity loading, which must be combined with effects of lateral loads for the complete design of the bents. The design of the spandrel girder is, therefore, deferred until after the discussion of lateral loads later in this section.

Design of the Concrete Columns

The general cases for the concrete columns are as follows (see Figure 16.52):

1. The interior column, carrying primarily only gravity loads due to the stiffened perimeter bents

2. The corner columns, carrying the ends of the spandrel beams and functioning as the ends of the perimeter bents in both directions

3. The intermediate columns on the north and south sides, carrying the ends of the interior girders and functioning as members of the perimeter bents

4. The intermediate columns on the east and west sides, carrying the ends of the column-line beams and functioning as members of the perimeter bents

Summations of the design loads for the columns may be done from the data given previously. Because all columns will be subjected to combinations of axial load and bending moments, these gravity loads represent only the axial compression action. Bending moments will be relatively

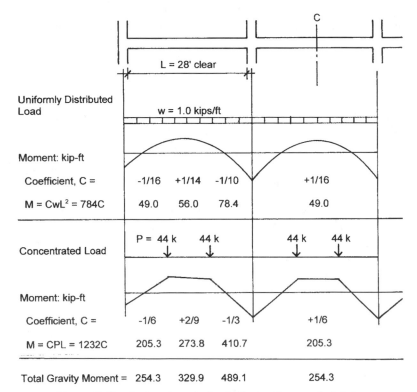

Figure 16.51   Factored gravity load effect on the spandrel girder.

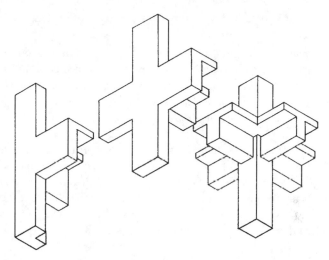

Figure 16.52　Relations between the columns and the floor framing.

low in magnitude on interior columns because they are framed into by beams on all sides. As discussed in Chapter 10, all columns are designed for a minimum amount of bending, so routine design, even when done for axial load alone, provides for some residual moment capacity. For an approximate design, therefore, it is reasonable to consider the interior columns for axial gravity loads only.

Figure 16.53 presents a summary of design for an interior column, using loads determined from a column load summation with the data given previously in this section. Note that a single size of 18-in. square is used for all three stories, a common practice permitting reuse of column forms for cost savings. Column load capacities indicated in Figure 16.53 were obtained from the graphs in Chapter 10.

A general cost-savings factor is the use of relatively low percentages of steel reinforcement. An economical column is, therefore, one with a minimum percentage (usually a threshold of 1% of the gross section) of reinforcement. However, other factors often affect design choices for columns; some common ones are as follows:

- Architectural planning of building interiors. Large columns are often difficult to plan around when developing of interior rooms, corridors, stair openings, and so on. Thus, the *smallest* feasible col-

umn sizes—obtained with maximum percentages of steel—are often
desired.

- Ultimate load response of lightly reinforced columns borders on
brittle fracture failure, whereas heavily reinforced columns tend to
have a yield-form of ultimate failure. The yield character is especially desirable for rigid frame actions in general, and particularly
for seismic loading conditions.

A general rule of practice in rigid frame design for lateral loadings
(wind or earthquakes) is to prefer a form of ultimate response described
as *strong column/weak beam failure*. In this example, this relates more to
the columns in the perimeter bents, but may also somewhat condition design choices for the interior columns because they will take *some* lateral
loads when the building as a whole deflects sideways.

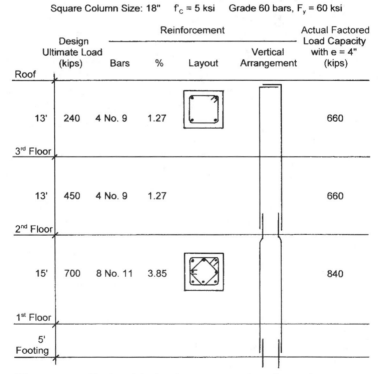

Figure 16.53   Design of the interior concrete column for gravity load only.

Column form may also be an issue that relates to architectural planning or to structural concerns. Round columns work well for some structural actions and may be quite economical for forming, but unless they are totally freestanding, they do not fit so well for planning of the rest of the building construction. Even square columns of large size may be difficult to plan around in some cases; an example is at the corners of stairwells and elevator shafts. T-shaped or L-shaped columns may be used in special situations.

Large bending moments in proportion to axial compression may also dictate some adjustment of column form or arrangement of reinforcement. When a column becomes essentially beamlike in its action, some of the practical considerations for beam design come into play. In this example, these concerns apply to the exterior columns to some degree.

For the intermediate exterior columns, there are four actions to consider:

1. The vertical compression due to gravity.
2. Bending moment induced by the interior framing that intersects the wall. These columns are what provides the end resisting moments shown in Figures 16.49 and 16.51.
3. Bending moments in the plane of the wall bent, induced by any unbalanced gravity load conditions (movable live loads) on the spandrels.
4. Bending moments in the plane of the wall bents due to lateral loads.

For the corner columns, the situation is similar to that for the intermediate exterior columns; that is, there is bending on both axes. Gravity loads will produce simultaneous bending on both axes, resulting in a net moment that is diagonal to the column. Lateral loads can cause the same effect because neither wind nor earthquakes will work neatly on the building's major axes, even though this is how design investigation is performed.

Further discussion of the exterior columns is presented in the following considerations for lateral load effects:

Design for Lateral Forces

The major lateral-force-resisting systems for this structure are as shown in Figure 16.54. In truth, other elements of the construction will also resist lateral distortion of the structure, but widening the exterior columns in the wall plane and using the very deep spandrel girders give these bents considerable stiffness.

Whenever lateral deformation occurs, the stiffer elements will attract the force first. Of course, the stiffest elements may not have the necessary strength and will, thus, fail structurally, passing the resistance off to other resisting elements. Glass tightly held in flexible window frames, stucco on light wood structural frames, lightweight concrete block walls, or plastered partitions on light metal partition frames may, thus, be fractured first in lateral movements (as they often are). For the successful design of this building, the detailing of the construction should be carefully done to ensure that these events do not occur, in spite of the relative stiffness of the perimeter bents. In any event, the bents shown in Figure 16.54 will be designed for the entire lateral load. They thus represent the safety assurance for the structure, if not a guarantee against loss of construction.

With the same building profile, the wind loads on this structure will be the same as those determined for the masonry structure in Section 16.15. As in the example in that section, the data given in Figure 16.44 are used to determine the horizontal forces on the bracing bent, as follows:

- $H_1 = 165.5(122)/2 = 10,096$ lb, say 10.1 kips/bent [44.9 kN/bent]
- $H_2 = 199.5(122)/2 = 12,170$ lb, say 12.2 kips/bent [54.3 kN/bent]
- $H_3 = 210(122)/2 = 12,810$ lb, say 12.8 kips/bent [56.9 kN/bent]

Figure 16.55a shows a profile of the north-south bent with these loads applied. For an approximate analysis, assume that the individual stories of the bent behave as shown in Figure 16.55b, with the columns developing an inflection point at their midheight points. Because the columns are all deflected the same sideways distance, the shear force in a single column may be assumed to be proportionate to the relative stiffness of the column. If the columns all have the same stiffness, the total load at each story for this bent would simply be divided by four to obtain the column shear forces.

Even if the columns are all the same size, however, they may not all have the same resistance to lateral deflection. The end columns in the bent are slightly less restrained at their ends (top and bottom) because they are framed on only one side by a beam. For this approximation, therefore, it is assumed that the relative stiffness of the end columns is one-half that of the intermediate columns. Thus, the shear force in the

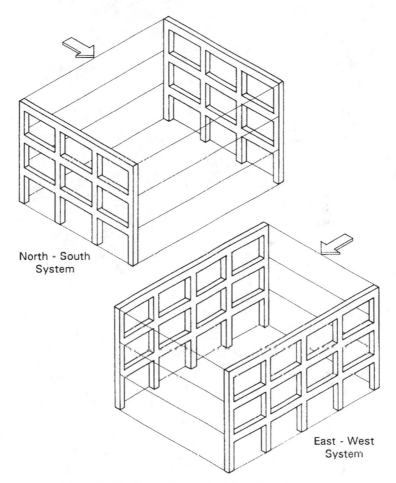

Figure 16.54   Form of the perimeter bent bracing system.

end columns is one-sixth of the total bent shear force, and the shear force in the intermediate column is one-third of the total force. The column shears for each of the three stories are, thus, as shown in Figure 16.55*c.*

The column shear forces produce bending moments in the columns. With the column inflection points (points of zero moment) assumed to be at midheight, the moment produced by a single shear force is simply the product of the force and half the column height. These column moments

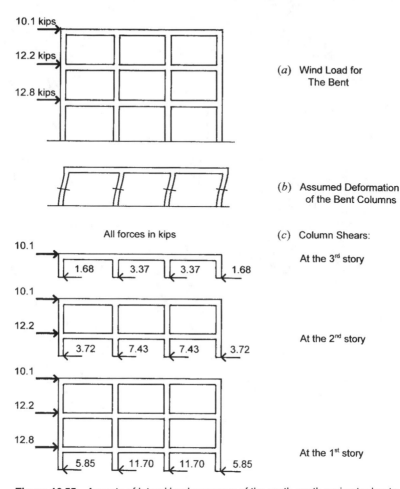

10.1 kips

12.2 kips

12.8 kips

(a) Wind Load for
The Bent

(b) Assumed Deformation
of the Bent Columns

All forces in kips

(c) Column Shears:

10.1

At the 3rd story

1.68   3.37   3.37   1.68

10.1

12.2

At the 2nd story

3.72   7.43   7.43   3.72

10.1

12.2

12.8

At the 1st story

5.85   11.70   11.70   5.85

Figure 16.55 Aspects of lateral load response of the north-south perimeter bents.

must be resisted by the end moments in the rigidly attached beams, and the actions are as shown in Figure 16.56. At each column/beam intersection, the sum of the column and beam moments must be balanced. Thus, the total of the beam moments may be equated to the total of the column moments, and the beam moments may be determined once the column moments are known.

For example, at the second-floor level of the intermediate column, the sum of the column moments from Figure 16.56 is

$$M = 48.3 + 87.8 = 136.1 \text{ kip-ft } [185 \text{ kN-m}]$$

Assuming the two beams framing the column to have equal stiffness at their ends, the beams will share this moment equally, and the end moment in each beam is thus

$$M = \frac{136.1}{2} = 68.05 \text{ kip-ft } [92.3 \text{ kN-m}]$$

as shown in the figure.

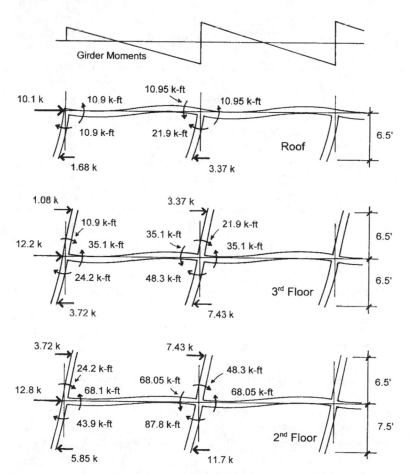

Figure 16.56 Investigation for column and girder bending moments in the north-south bents.

The data displayed in Figure 16.56 may now be combined with that obtained from gravity load analyses for a combined load investigation and the final design of the bent members.

Design of the Bent Columns

For the bent columns, the axial compression due to gravity must first be combined with any moments induced by gravity, for a gravity-only analysis. Then the gravity load actions are combined with the results from the lateral force analysis, using the usual adjustments for this combined loading. In the strength method, this occurs automatically through the use of the various load factors for the different loadings. In the stress method, the combined gravity and lateral load condition uses an adjusted stress—in this case, one increased by one-third. Because our illustration uses strength methods, a comparison can be made between the effects due to gravity only and those due to gravity plus lateral loads. This is accomplished by using the load factors for the two combinations.

Gravity-induced moments for the girders are taken from the girder analysis in Figure 16.51 and are assumed to produce column moments as shown in Figure 16.57. The summary of design conditions for the corner and intermediate columns is given in Figure 16.58. The dual requirements for the columns are given in the bottom two lines of the table in Figure 16.58.

When bending moment is very high in comparison to the axial load (very large eccentricity), an effective approximate column design can be determined by designing a section simply as a beam with tension reinforcement; the reinforcement is then merely duplicated on both sides of the column.

Design of the Bent Girders

The spandrel girders must be designed for the same two basic load conditions as discussed for the columns. The summary of bending moments for the third-floor spandrel girder is shown in Figure 16.59. Values for the gravity moments are taken from Figure 16.51. Moment induced by wind is that shown in Figure 16.56. It may be noted from the data in Figure 16.59 that the effects of gravity loading prevail and that wind loading is not a critical concern for the girder. This would most likely not be the

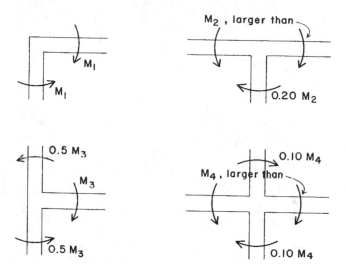

Figure 16.57 Assumptions for approximations of the distribution of bending moments in the bent columns due to gravity loading.

case in lower stories of a much taller building or possibly with a combined loading that includes major seismic effects.

Figure 16.60 presents a summary of design considerations for the third-floor spandrel girder. The construction assumed here is the one shown in Figure 16.48, with the very deep, exposed girder. Some attention should be given to the relative stiffness of the columns and girders, as discussed in Section 11.4. Keep in mind, however, that the girder is almost three times as long as the column and, thus, may have a considerably stiffer section without causing a disproportionate relationship to occur.

For computation of the required flexural reinforcement, the T-beam effect is ignored, and an effective depth of 40 in. is assumed. Required areas of reinforcement may, thus, be derived as

$$A_s = \frac{M}{\phi f_y jd} = \frac{M \times 12}{0.9(40 \times 0.9 \times 40)} = 0.00926M$$

Values determined for the various critical locations are shown in Figure 16.60. It is reasonable to consider the stacking of bars in two layers

Note: Axial loads in kips, moments in kip-ft, dimensions in inches.

Story	Third	Second	First
Gravity Load Only:			
Axial Live Load	35	67	103
Axial Dead Load	83	154	240
Live Load Moment	32	38	38
Dead Load Moment	128	90	90
Ultimate Gravity Load: (1.2D + 1.6L)			
Axial Load	100+56 = 156	185+107 = 292	288+165 = 453
Moment	154+51 = 205	108+61 = 169	108+61 = 169
e	15.8 in.	7.0 in.	4.5 in.
Combined with Wind:			
Wind Load Moment	21.9	48.3	87.8
Ultimate Moment with Wind (1.2D+1.6W+L)	154+35+32 = 221	108+77+38 = 223	108+141+38 = 287
Ultimate Axial Load with Wind (1.2D + L)	100+35 = 135	185+67 = 252	288+103 = 391
e	19.6 in.	10.6 in.	8.8 in.
Choice of Column from Figure 10.11	?? Design as a beam for M = 221	14 × 20 6 No. 9	14 × 20 6 No. 9

3rd story: Use 14×24, d = 21 in., $A_s = M_u/\phi f_y jd = (221×12)/0.9[60(0.9×21)] = 2.6$ in.^{2}
Use same as 2nd story, 6 No. 9 bars (3 at each end). Use 14×24 for all stories.

Figure 16.58 Design of the north-south bent columns for combined gravity and lateral loading.

in such a deep section, but it is not necessary for the selection of re-inforcement shown in the figure.

The very deep and relatively thin spandrel should be treated somewhat as a wall/slab, and, thus, the section in Figure 16.60 shows some additional horizontal bars at midheight points. In addition, the stirrups shown should be of a closed form (see Figure 7.5) to serve also as ties, as vertical reinforcement for the wall/slab, and (with the extended top) as negative moment flexural reinforcement for the adjoining slab. In this situation, it would be advisable to use continuous stirrups at a maximum

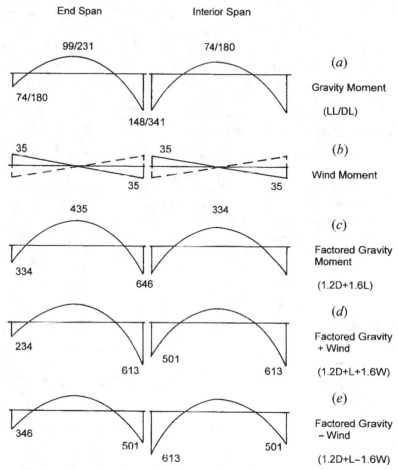

Figure 16.59 Combined gravity and lateral bending moments for the spandrel girders.

spacing of 18 in. or so for the entire girder span. Closer spacing may be necessary near the supports if the end shear forces require it.

It is also advisable to use some continuous top and bottom reinforcement in spandrels. This relates to some of the following possible considerations:

- In case of miscalculation of lateral effects, this will give some reserved, reversal bending capacity to the girders.

- A general capability for torsional resistance throughout the beam length. (Intersecting beams produce this effect.)
- There will be something there to hold up the continuous stirrups.
- Some reduction of long-term creep deflection will be provided, with all sections doubly reinforced. This helps to keep load off the window mullions and glazing.

Design of the Foundations

Unless site conditions require the use of a more complex foundation system, it is reasonable to consider the use of simple shallow bearing foundations (footings) for Building Five. Column loads will vary, depending on which of the preceding structural schemes is selected. The heaviest loads are likely to come with the all-concrete structure.

The most direct solution for concentrated column loads is a square footing, as described in Section 13.6. A range of sizes of these footings is given

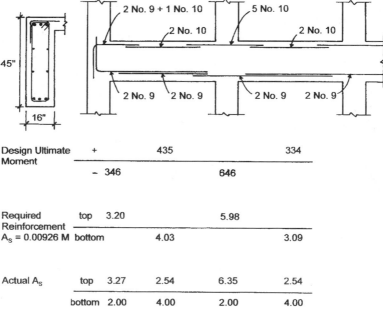

Figure 16.60 Design of the spandrel girder for the combined gravity and lateral loading effects.

in Table 13.5. For a freestanding column, the choice is relatively simple once an acceptable design pressure for the supporting soil is established.

Problems arise when the conditions at the base of a column involve something other than a freestanding case for the column. In fact, this is the case for most of the columns in Building Three. Consider the structural plan as indicated in Figure 16.46. All but two of the interior columns are adjacent to construction for the stair towers or the elevator shaft.

For the three-story building, the stair tower may not be a problem, although in some buildings these are built as heavy masonry or concrete towers and are used for part of the lateral bracing system. This might possibly be the case for the structure in Section 16.11.

Assuming that the elevator serves the lowest occupied level in the building, there will be a considerably deep construction below this level to house the elevator pit. If the interior footings are quite large, they may come very close to the elevator pit construction. In this case, the bottoms of these footings would need to be dropped to a level close to that of the bottom of the elevator pit. If the plan layout results in a column right at the edge of the elevator shaft, this is a more complicated problem because the elevator pit would need to be on top of the column footing.

For the exterior columns at the building edge, there are two special concerns. The first has to do with the necessary support for the exterior building wall, which coincides with the column locations. If there is no basement and the exterior building wall is quite light in weight—possibly a metal curtain wall that is supported at each upper level—the exterior columns will likely get their own individual square footings, and the wall will get a minor strip footing between the column footings. This scheme is shown in the partial foundation plan in Figure 16.61a. Near the columns, the light wall will simply be supported by the column footings.

If the wall is very heavy, the solution in Figure 16.61a may be less feasible, and it may be reasonable to consider the use of a wide strip footing that supports both the wall and the columns as shown in Figure 16.61b. The ability of the wall to serve as a distributing member for the uniform pressure on the strip footing must be considered.

The other concern has to do with the presence or absence of a basement. If there is no basement, it may be theoretically possible to place footings quite close to the ground surface, with a minimal penetration of the general foundation construction below grade. If there is a basement, there will likely be a reasonably tall concrete wall at the building edge. Regardless of the wall construction above grade, it is reasonable to consider the use of the basement wall as a distributing member for the col-

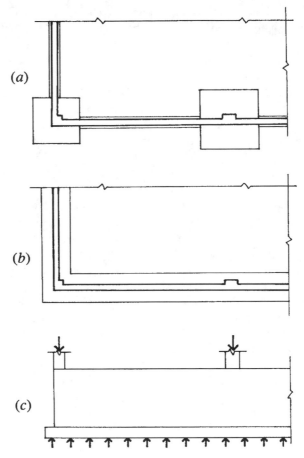

Figure 16.61   Building Five, considerations for the foundations: (*a*) partial plan with individual column footings, (*b*) partial plan with continuous wall footing, and (*c*) use of a tall basement wall as a distribution girder.

umn loads, as shown in Figure 16.61*c*. Again, this is only feasible for a low-rise building, with relatively modest loads on the exterior columns.

16.17   BUILDING FIVE: ALTERNATIVE FLOOR STRUCTURE

Figure 16.62 shows a partial plan and some details for a concrete flat slab system for the roof and floor structures for Building Five. The general na-

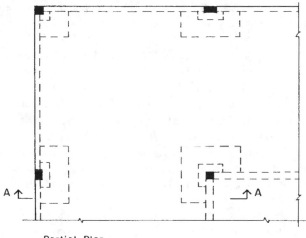

Partial Plan

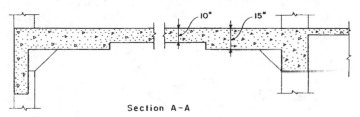

Section A-A

Figure 16.62 Building Five: alternative structure for the upper floor, using a concrete flat slab.

ture of this type of system is discussed in Section 9.3. Features of the system, as shown in Fig. 16.62, include the following:

1. A general, solid, 10-in.-thick slab without beams in the major portion of the structure outside the building core.
2. A thickened portion, called a *drop panel,* around the supporting columns, ordinarily extending to one-sixth the span from the column on all sides. The thickness increase shown is one-half of the general slab thickness.
3. A *column capital,* consisting of a truncated, inverted pyramidal form.

4. Use of the same spandrel beam and the same slab-and-beam core framing as in the example in Section 16.16.

The flat slab is generally more feasible when live loads are very high and there is a considerable number of continuous bays of the structure in both directions. The system for this building is marginally feasible and might be justified on the basis of some other building design considerations. For example, the story height may be decreased because downward-protruding beams do not interfere with air ducts, sprinkler piping, and so on, in the main portion of the floor. This could mean a building several feet shorter, with resultant savings in curtain walls, columns and partitions, stairs, elevator housing, piping and wiring risers, and so on.

For some occupancies, in fact, the drop ceiling construction might be eliminated, with all overhead items exposed beneath the considerably simpler form of the concrete structure. This is a real advantage for parking garages and industrial buildings, but not so popular with office buildings.

Tabular data for the design of this system are given in the *CRSI Handbook* (Reference 5). The system is highly indeterminate and a bit complicated for reinforcement. However, it has been used extensively since its development in the early twentieth century, and routine design of common examples is pretty well "canned" for repetition by now. See the general discussion of two-way systems in Chapter 9.

GLOSSARY

Every topic area or field has its own private language. To help readers with some of the materials in this book, as well as with that in other publications about concrete, the following brief glossary of terms is provided. It is generally limited to terms that are "insider talk" in the concrete design/build business.

Admixture. In general, anything added to the wet concrete mix besides the basic elements of water, cement, and aggregate; usually done to modify the wet mix or the finished concrete in some manner.

Aggregate. Inert, loose material that makes up the largest part (typically $\frac{2}{3}$ to $\frac{3}{4}$) of the bulk of concrete; what the water and cement paste holds together; ordinarily consists of stone—ranging in size from medium fine sand to coarse gravel.

Air-Entrainment. Captive air (tiny bubbles) within the concrete mass, produced deliberately to alter the character of the concrete.

Backfill. General description of soil placed back into an excavation, often up against some constructed structure, such as a basement wall.

Cap (Column). Enlarged top of a column, used to shorten the span between columns, increase shear periphery, or reduce negative bending moments in the spanning structure; or all of the above.

Cold Joint. Point within a continuous concrete structure where one casting is stopped and another begun later, but only after the first casting is allowed to harden (get cold).

Cylinder (Test). A sample of a batch of wet concrete made in the form of a cylinder; for use in a compression test on the hardened concrete.

Doubly Reinforced. Concrete member with both tension and compression reinforcement, usually opposed in bending.

Dowel. Steel reinforcing rod usually extending from one cast member for subsequent encasement in an adjacent casting of concrete or adjacent construction of masonry.

Drop (Form). Forming for an element that consists of an extension of the concrete mass beneath a slab structure; usually refers to flat slab construction although general forming of beams for slab and beam construction is technically drop forming.

Factored Load. Service load multiplied by a load factor to adjust for use in strength design.

Grade. Elevation of the top of the ground at a point; the slope of a non-flat surface of concrete or earth.

Insert. Something besides the steel reinforcement that is inserted into the forms to be cast into the concrete; for example, anchor bolts.

Jobsite. Where the construction is happening.

Plain Concrete. Concrete cast without reinforcement or prestressing.

Pour. Act of depositing wet concrete into forms; the cast portion completed in a single continuous operation.

Pour Joint. See *Cold Joint.*

Poured-in-Place. Concrete cast where it is intended to stay; also called *sitecast.*

Precast. General description of a concrete element not cast where it is intended to be used. May be cast off the site and transported in or cast elsewhere on the site and moved into place.

Rebar. Slang for *reinforcing bar.*

Resistance Factor. Factor for adjustment of the ultimate resistance of a structural element for use in strength design.

Sack. A unit of dry cement, consisting of one cubic foot or 94 lbs. For small volume use may actually be packed in a sack.

Sitecast. See *Poured-in-Place.*

Slab. A horizontal, planar element of concrete, whether deposited on the ground (as a paving slab) or cast in midair (for a supported slab).

Slump (Test). Old-fashioned method for determining the water content of a wet mix of concrete, by observing the amount of sag (slump) of a batch sample that is formed and then suddenly released from the form. The dimension of slump of the released mass is an index of both workability and strength of the concrete.

Structural Concrete. Term now used by the ACI Code for general reference, in place of the particular terms of *reinforced concrete* or *prestressed concrete.*

Tie (Wire). Soft steel wire carried by persons engaged in placing the steel bars in the forms. Intersecting bars are typically tied to hold the general set of bars more firmly in place during casting of the concrete.

Truck. As in a "truck of concrete," referring to a transit mixer truck used to deliver wet concrete to the jobsite. A larger unit than *yard* for reference since a truck usually carries five or more yards of concrete.

Unreinforced. Ungrammatical term commonly used to describe concrete or masonry structures without reinforcement. Unreinforced concrete is also called *plain concrete.*

Waffle. Two-way spanning concrete joist construction that looks like a giant waffle when viewed from below.

Wet. As in "wet concrete"; refers to the concrete fresh from the mixer, in a semifluid state before hardening occurs.

Yard. Short for 1 cubic yard or 9 cubic feet; the usual volumetric measure for quantities of concrete.

STUDY AIDS

The materials in this section are provided for readers to use in order to test their general understanding of the book presentations. It is recommended that upon completion of reading of an individual chapter, readers use the materials here as a review.

TERMS

Using the text of the chapter indicated, together with the Index and Glossary, review the definitions and significance of the following terms.

Chapter 1

Concrete

Cement

Portland cement

Sitecast concrete

Precast concrete

Prestressed concrete

Structural concrete

Chapter 2

Aggregate
High-early-strength cement
  (or concrete)
Air-entrained concrete
Admixture
Specified compressive strength
Modulus of elasticity
Creep
Modulus ratio
Workability
Cover
Compression test
Slump test
Reinforcement
Prestressing
Pretensioning
Post-tensioning
Tilt-up wall
Lightweight concrete
Insulating concrete
Fiber-reinforced concrete

Chapter 3

Construction joint
Control joint
Cold joint
Minimum reinforcement
Shrinkage and temperature
  reinforcement
Spacing requirements for
  reinforcement

Chapter 4

Working stress method
  (stress method)
Service load
Strength method
Factored load
Continuous beam
Rigid frame
Moment-resistive joint
  (connection)
Free body diagram
Deformed shape
Sidesway

Chapter 5

LRFD
Load factor
Strength reduction factor
  (resistance factor)

Chapter 6

Effective depth of a beam
Balanced section
Under-reinforced section
Design strength (strength method)
Rectangular stress block
T-beam
Doubly reinforced beam

Chapter 7

Punching shear
Diagonal tension stress
Stirrup

Chapter 8

Bond stress
Development length
Standard hook
Reinforcing dowel
Lapped splice

Chapter 9

One-way slab and beam system
Two-way spanning system
Concrete joist construction
Waffle construction
Flat slab
Composite construction

Chapter 10

Column interaction response
Tied column
Spiral column
P-delta effect

Chapter 11

Rigid frame
Bent
Lateral load
Sidesway

Chapter 12

Bearing wall
Retaining wall
Shear wall
Freestanding wall
Grade wall

Chapter 13

Shallow bearing foundation
Footing
Site survey
Allowable bearing pressure
Active lateral pressure
Passive lateral pressure
Presumptive soil properties
Wall footing
Column footing
Rectangular footing
Combined footing
Pedestal
Deep foundation
End-bearing pile
Friction pile
Belled pier (caisson)
Abutment

Chapter 14

Slab on grade
Subgrade
Framed floor on grade
Cantilever retaining wall

Chapter 15

Dead load
Live load
Building code
Live load reduction
Load periphery

QUESTIONS

Note: Answers to these questions follow the last question.

Chapters 1 to 5

1. What is the primary structural limitation of concrete that generates the need for reinforcement?
2. What is the significance of having a well-graded range of sizes for the aggregate in concrete?
3. What is the primary factor that affects the unit density (weight) of structural concrete?
4. What is the purpose of the surface deformations on steel reinforcing bars?
5. What structural property is most significant for the steel used for concrete reinforcement?
6. During the curing period for concrete (after casting and before major structural use), what significant controls should be exercised?
7. What are the significant concerns that establish cover requirements for reinforcement in concrete construction?
8. With regard to the assumption of stress conditions, what is the difference between the stress and strength methods of design for reinforced concrete?

Chapters 6 to 8

1. What internal force development is represented by the rectangular stress block in strength investigation of a reinforced concrete beam?
2. Other than spacing limitations and bar diameters, what factors establish the maximum number of bars that can be placed in a single layer in a reinforced concrete beam?
3. Why is compressive strength of the concrete generally not critical for T-beam actions in sitecast construction?
4. For shear actions in concrete beams, what acts together with the vertical stirrups to resist diagonal tension stresses?
5. Stirrups are designed to resist what portion of the total shear force in a concrete beam?

6. Why is development length usually less critical for small-diameter reinforcing bars?

7. Why are longer development lengths required for bars of higher-grade steel?

8. In what form is the anchorage capacity of a hook expressed?

9. What is the basis for the establishment of the required length for a lapped splice?

Chapter 9

1. What are the usual considerations for determination of slab thickness in a slab and beam system?

2. How is the minimum depth required for slabs and beams affected by span and support conditions?

3. What is the essential difference in structural action between concrete joist construction and waffle construction?

4. What structural improvements are achieved by the use of column capitals and drop panels in flat slab construction?

5. What is the essential function of the shear developers (welded steel studs) in composite construction with steel beams and a sitecast concrete slab?

Chapter 10

1. With regard to effects on the vertical reinforcement, what is the primary purpose of the ties in a tied column?

2. With a column subjected to a large bending moment, why is a larger moment possible with the addition of a minor axial compression load?

3. What is the usual basis for limitation of the number of bars that can be placed in a spiral column?

Chapter 11

1. Why do sitecast concrete framing systems ordinarily constitute rigid frame structures?

2. Why is it necessary to coordinate the selection of reinforcement in all intersecting members at a single joint in sitecast concrete frame construction?

3. When structural frames and structural walls interact in resisting lateral loads, why do the walls tend to take a major portion of the lateral load?

4. Why is wind sometimes not a critical concern for design of individual structural members even though it is an addition to gravity loadings?

5. With regard to the form of deformation of a rigid frame, what is the significance of the relative stiffness of the frame members?

Chapter 12

1. When a concrete basement wall is reinforced with a single layer of vertical bars, why are the bars placed nearer to the inside face of the wall?

2. What is the purpose of the vertically projected portion that is sometimes used beneath the footing of a cantilever retaining wall?

Chapter 13

1. What are the principal engineering properties of soils that most affect the design of building foundations?

2. Other than practical concerns for forming of the cast concrete, what favors the use of a square plan shape for a column footing?

3. What is the purpose of the longitudinal reinforcement in a wall footing?

4. How does the use of a pedestal create the possibility for a thinner footing?

Chapter 14

1. Besides the thickness of the concrete slab, what are the significant factors that determine the load-carrying strength of a concrete paving slab?

2. What is the usual condition that requires the use of a framed floor on grade?

3. How does a cantilever retaining wall resist horizontal sliding?

4. What is the usual combination of loads on an abutment support?

Chapter 15

1. Of what does the design dead load primarily consist?
2. What primary factor affects the percentage of live load reduction?
3. In what unit is the load periphery for a column expressed?
4. Why is the achievement of optimal structural efficiency not always a dominant concern for the general cost of building construction?

ANSWERS TO QUESTIONS

Chapters 1 to 5

1. Low resistance to tension, as generated mostly by bending and shear actions.
2. The aggregate will pack into the most dense mass, with smaller pieces filling the void between larger pieces.
3. The unit density of the coarse aggregate (gravel in ordinary concrete), which makes up the major portion of the concrete volume.
4. To develop a mechanical lock (grip) between the bar surface and the surrounding concrete.
5. The yield strength and range of ductility.
6. Temperature, moisture content, and lack of demand for high stress resistance.
7. Size of the concrete member, fire resistance requirements, and exposure conditions to weather or to contact with soil.
8. The stress method uses service (actual, working) level stress conditions. The strength method uses only failure (limit state) conditions.

Chapters 6 to 8

1. Development of the internal compression force in the concrete that opposes the tension in the reinforcement to develop bending resistance.
2. Beam width, presence and size of stirrups, size of the largest aggregate, and general code requirements.
3. Typical size of slabs (thickness) and beams (primarily depth) result in an extensive cross-sectional area of concrete for the develop-

ment of compression necessary to oppose the tension in practical amounts of reinforcement.

4. The horizontal, tension-resisting bars near the beam ends.
5. The shear in excess of that permitted to be resisted by the concrete.
6. Development occurs on the bar's surface as the steel attempts to generate tensile stress on the bar's cross-sectional area. The smaller the bar, the greater the ratio of surface area to cross-sectional area, and, thus, the more surface available in proportion to the bar's tension strength.
7. Required development relates to potential bar tensile strength, which is greater if the usable stress level is higher.
8. In units of equivalent bar length.
9. The development length required for the bars being spliced.

Chapter 9

1. Span of the slab, design loading, fire resistance requirements, and, perhaps, T-beam action of the framing members.
2. These form the basis for the recommended minimum thicknesses in the ACI Code. Deflection is the primary issue.
3. Joists form a one-way spanning system; the waffle is a two-way spanning system.
4. The clear span of the slab is slightly reduced, and the bending and shear resistances near the column are increased.
5. To make the steel beams and the concrete slab work together as a single unit to resisting bending. This is equivalent to the use of the surface deformations on reinforcing bars.

Chapter 10

1. To prevent the slender bars from buckling and bursting out through the thin cover of concrete.
2. Code-imposed design requirements result in columns in which tension yielding of the vertical bars is the initial failure condition in bending. Adding a small amount of compression produces a small compression, prestressed condition in the steel, permitting the development of additional tension resistance. Eventually, however, added compression will result in column action failure unless the bending moment is reduced.
3. Spacing requirements for the bars in the circular arrangement.

Chapter 11

1. Simultaneous casting of members and/or extension of reinforcing through the frame joints produce moment-resistive connections between members.
2. Two reasons: extending bars through joints from member-to-member means sharing of single bars, and the traffic of intersecting bars must be feasible for the joint.
3. They are much greater in stiffness (deformation resistance) for shear force in their own planes.
4. Different load factors (strength method) or increased allowable stresses (stress method) permit some amount of "free" wind loading before it becomes critical.
5. If some frame members are considerably greater in relative stiffness, the normal interactive flexing of the frame members will be affected. Very stiff columns or beams may scarcely inflect at all.

Chapter 12

1. Tension develops on the inside of the wall as the wall spans vertically in resisting the soil pressure on the outside surface.
2. Enhancement of the resistance of the wall to horizontal sliding.

Chapter 13

1. Size of solid particles, range of size of particles, in-place density, relative hardness (penetration resistance), water content, potential instability, and presence of organic materials.
2. Ease and simplicity of placement of the two intersecting layers of reinforcement.
3. To resist cracking due to shrinkage stresses and, perhaps, to help the footing to act as a grade beam for spanning across uneven soil conditions.
4. By reducing the magnitude of shear and bending in the footing.

Chapter 14

1. Use of reinforcement (steel or fiber) and quality of the supporting subbase.
2. Lack of reliability of the supporting soil.

3. By a combination of horizontal soil pressure on the footing, soil friction on the footing bottom, and—if used—a shear key.

4. Vertical and horizontal forces.

Chapter 15

1. Weight of the building construction.

2. Area of the loaded surface carried by the member being considered.

3. In square feet of supported surface area.

4. Structural efficiency may well be less important than the effect of the structure on cost of the general construction.

ANSWERS TO
EXERCISE PROBLEMS

Chapter 4

4.6.A. $R = 10$ kips up and 110 kip-ft counterclockwise.

4.6.B. $R = 5$ kips up and 24 kip-ft counterclockwise.

4.6.C. $R - 6$ kips to the left and 72 kip ft counterclockwise.

4.6.D.   Left $R = 4.5$ kips up, right $R = 4.5$ kips down and 12 kips to the right.

4.6.E.   Left $R = 4.5$ kips down and 6 kips to the left; right $R = 4.5$ kips up and 6 kips to the left.

Chapter 6

6.3.A.   Possible choice: $b = 12$ in. [305 mm], $d = 14.5$ in. [368 mm], requires 7.60 in.^{2} [4903 mm^2], use 5 No. 11 bars. However, width required to get bars into one layer is critical; must use wider beam or two layers of bars.

6.3.B.   Possible choice: $b = 12$ in. [305 mm], $d = 10.76$ in. [273 mm], requires 4.02 in.^{2} [2594 mm^2], use 4 No. 9, must use wider beam or two layers of bars.

6.3.C.   From work for Problem 6.3.A, this section is larger than a balanced

415

section, try $a/d = 0.4$, required area is 2.81 in.^{2} [1813 mm^2], actual $a/d = 0.086$, area is 2.34 in.^{2} [1510 mm^2], use 3 No. 8, width OK.

6.3.D. Required area = 1.46 in.^{2} [942 mm^2], use 3 No. 7.

6.3.E. Possible choices are: 12 × 20, 2 No. 10 + 2 No. 11; 12 × 24, 4 No. 10; 15 × 20, 4 No. 11; 15 × 25, 4 No. 9; 18 × 24, 5 No. 9; 18 × 36, 3 No. 10; 20 × 30, 3 No. 10.

6.3.F. Possible choices are: 18 × 30, 8 No. 11; 18 × 36, 7 No. 11; 20 × 30, 9 No. 11; 20 × 35, 7 No. 11; 20 × 40, 6 No. 11; 24 × 32, 8 No. 11; 24 × 40, 7 No. 11; 24 × 48, 5 No. 10.

6.5.A. For bending moment approximate required area is 3.84 in.^{2} [2478 mm^2]; however, minimum based on width of flange is 5.52 in.^{2} [3562 mm^2].

6.5.B. Flexure requires 3.77 in.^{2} [2432 mm^2], but minimum is 5.46 in.^{2} [3523 mm^2].

6.6.A. Section does not need compressive reinforcement, but if desired use 5 No. 9 for tension and 3 No. 7 for compression.

6.6.B. Section does not need compressive reinforcement, but if desired use 6 No. 11 for tension and 3 No. 9 for compression.

6.6.C. Section does not need compressive reinforcement, but if desired use 5 No. 11 for tension (width required is 17 in.) and 3 No. 9 for compression.

6.6.D. Section does not need compressive reinforcement, but if desired use 8 No. 11 for tension (requires greater width or two layers) and 4 No. 9 for compression.

6.7.A. For deflection use 8-in. slab, reinforce with No. 8 at 16 in., No. 7 at 12 in., No. 6 at 9 in., or No. 5 at 6 in.; use No. 4 at 12 in. for temperature.

6.7.B. Requires 11-in. slab, No. 5 at 11 in., No. 6 at 16, or No. 7 at 18 in.; use No. 4 at 9 for temperature.

Chapter 7

7.4.A. Possible choice: 1 at 6; 8 at 13.

7.4.B. Minimum reinforcement: 1 at 5; 8 at 11.

7.4.C. Possible choice: 1 at 6; 4 at 13.

7.4.D. Possible choice: 1 at 5; 9 at 11.

Chapter 8

8.1.A. Required length in cantilever is 17 in., 34 in. provided; required length in support is 13 in., 22 in. provided. Anchorage as shown is adequate, but use hook anyway.

8.1.B. Lengths provided are adequate.

8.2.A. Required length is 4.9 in., but use full available length.

8.2.B. Required length is 5.6 in., but use full available length.

Chapter 9

9.1.A.   For deflection need 5.5-in. slab, referring to Figure 9.4, left to right, with all No. 4 bars, use spacings of 16, 18, 13, 18, 16 in.

9.1.B.   For deflection need 5.0-in. slab, referring to Figure 9.4, left to right, with all No. 4 bars, use spacings of 15, 15, 12, 15, 15 in.

Chapter 10

10.10.A.   12 in. square, 4 No. 8.

10.10.B.   14 in. square, 8 No. 9.

10.10.C.   18 in. square, 8 No. 11.

10.10.D.   12 × 16, 6 No. 7.

10.10.E.   12 × 16, 6 No. 7.

10.10.F.   14 × 20, 6 No. 9.

10.11.A.   14-in. diameter, 4 No. 11.

10.11.B.   16-in. diameter, 6 No. 10.

10.11.C.   20-in. diameter, 8 No. 11.

Chapter 12

12.2.A.   Required wall resistance is 80 kips [356 kN], minimum wall thickness is 7.2 in. [183 mm], use 8-in. [203 mm], bearing capacity is 153 kips [680 kN]—not critical, column capacity is 264 kips [1174 kN]—not critical, for reinforcement use No. 4 at 16 in. vertical and No. 5 at 15 in. horizontal.

12.2.B.   Required wall resistance is 103 kips [458 kN], bearing capacity is 323 kips [1437 kN]—not critical, column capacity is 744 kips [3309 kN]—not critical, for reinforcement use No. 4 at 13 in. vertical and No. 5 at 12 in. horizontal.

12.3.A.   For 12-in. [305 mm] wall, minimum vertical reinforcement is 0.216 in.^{2}/ft [457 mm^2/m], for flexure use 0.256 in.^{2}/ft [542 mm^2/m], use No. 5 at 14 in. vertical and No. 5 at 10 in. horizontal. For 14-in. wall, required area is 0.215 in.^{2}/ft [455 mm^2/m], use No. 5 at 13 in., for horizontal use No. 6 at 12 in.

12.3.B.   For 15-in. wall, use No. 5 at 10 in. vertical and No. 6 at 11 in. horizontal.

Chapter 13

13.5.A.   Possible choice is 6-ft-9-in. wide by 17 in. thick footing with No. 5 at 12 in. in short direction and 7 No. 5 in long direction.

13.5.B.   Possible choice is 5-ft-wide by 16-in.-thick footing with No. 5 at 18 in. in short direction and 5 No. 5 in long direction.

13.6.A.   From Table 13.6 a possible choice is 9-ft-square by 21-in.-thick footing with 9 No. 8 each way; computations will permit 8 ft 8 in. wide by 20 in. thick with 7 No. 8 each way.

13.6.B.   From Table 13.6 a possible choice is a 12-ft-square by 32-in.-thick footing with 11 No. 10 each way; computations will permit 11-ft-9-in. square by 30-in.-thick footing with 10 No. 10 bars each way.

13.8.A.   Factored load is 368 kips [1637 kN], minimum height for development is 31 in. [787 mm], recommended minimum width is 18 + 12 = 30 in. [762 mm], bearing capacity is 758 kips [3372 kN]—OK, column capacity is 890 kips [3959 kN]—OK, reinforcement is not required, but some would be helpful.

13.8.B.   Factored load is 664 kips [2953 kN], minimum height for development is 38 in. [965 mm], recommended width is 22 + 12 = 34 in. [864 mm], bearing capacity is 1049 kips [4666 kN]—OK, column capacity is 1143 kips [5084 kN]—OK, reinforcement is not required, but some would be helpful.

Chapter 14

14.3.A.   Maximum soil pressure is 981 psf [47.0 kPa]—less than 2000 psf [95.8 kPa], OK; safety factor against overturn is 2.53—more than 1.5, OK; vertical reinforcement in wall is No. 3 at 18 in.

14.3.B.   Maximum soil pressure is 818 psf [39.2 kPa]—less than 2000 psf [95.8 kPa], OK; safety factor against overturn is 3.59—greater than 1.5, OK; vertical reinforcement in wall is No. 3 at 16 in. or No. 4 at 18 in.

REFERENCES

1. *Building Code Requirements for Structural Concrete,* ACI 318–05 and Commentary, ACI 318R-05, American Concrete Institute, Detroit, MI, 2005.

2. *Minimum Design Loads for Buildings and Other Structures,* SEI/ASCE 7–02, American Society of Civil Engineers, Reston, VA, 2003.

3. *Uniform Building Code, Volume 2: Structural Engineering Provisions,* International Conference of Building Officials, Whittier, CA, 1997.

4. Jack McCormac and James Nelson, *Design of Reinforced Concrete,* 6th ed., Wiley, Hoboken, NJ, 2005.

5. *CRSI Design Handbook 2002,* Concrete Reinforcing Steel Institute, Schaumburg, IL, 2002.

6. James Ambrose and Dimitry Vergun, *Simplified Building Design for Wind and Earthquake Forces,* 3rd ed., Wiley, Hoboken, NJ, 1995.

7. James Ambrose, *Simplified Design of Building Foundations,* 2nd ed., Wiley, Hoboken, NJ, 1988.

8. James Ambrose, *Simplified Mechanics and Strength of Materials,* 6th ed., Wiley, Hoboken, NJ, 2005.

INDEX